Practical Application

of **SPC** in the

Wire & Cable Industry

Practical Application

of **SPC** in the

Wire & Cable Industry

DOUGLAS B. RELYEA

Quality Resources
A Division of The Kraus Organization Limited
White Plains, New York

Printed in the United States of America

95 94 93 92 91   10 9 8 7 6 5 4 3 2

Quality Resources
A Division of The Kraus Organization Limited
One Water Street
White Plains, NY 10601

Library of Congress Cataloging-in-Publication Data

Relyea, Douglas B.
    Practical application of SPC in the wire & cable industry /
Douglas B. Relyea.
      p. cm.
    ISBN 0-527-91643-9
    1. Process control—Statistical methods. 2. Electric wire and
cable industry—Quality control. 3. Wire industry—Quality control.
4. Wire-rope industry—Quality control. I. Title.
TS156.8.R44   1990
670.42—dc20   89-70237
                                                             CIP

This book is dedicated to Kathleen,
who has helped and supported me
in my consulting and writing career.

Contents

Figures

Tables

Practical
Application

of **SPC** in the

Wire & Cable
Industry

1

Proof of the Need

THE DECLINE OF MANUFACTURING IN THE UNITED STATES

In July of 1988, 29 percent of the wire companies queried by *Wire Industry News*[1] responded that they found imports to be a serious problem. Other comments ranged from complaints of foreign "dumping" of products, to the shortage of domestically produced hot-rolled rod.

In approaching the 1990s, the wire and cable industries, like many other American businesses, are clearly feeling the effects of offshore competition. Every day, magazine articles, newspaper editorials, and talk shows address the apparent lack of ability of American industry to compete in the world market.

Some giants in American industry have surrendered their ability to manufacture some products and have become distributors of goods made overseas. For example, Kodak and RCA market video cameras and recorders made in Japan; General Electric places its name on microwave ovens manufactured in Asia; Chrysler, Ford, and General Motors import small cars made in Japan for resale in the United States.

In 1979, there were 26.5 million Americans employed in manufacturing. That figure dropped to 25 million in 1986,[2] leaving one and a half million Americans no longer producing goods for resale—one and a half million Americans either idle or employed in a service industry such as restaurants or insurance sales. And the decline in manufacturing jobs continues; over 100,000 were lost in September and October of 1989 alone.

The impact of American companies giving up on manufacturing goes beyond the problem of job losses. There are far-ranging effects that will be felt by every segment of our society—effects that may not be fully comprehended for years to come.

One side effect of giving up on manufacturing is found in the truism that an organization loses interest in the development of a product it no longer manufactures.

This happened at General Electric.[3,4] After years of trying to produce high-quality televisions at competitive prices, GE gave up, and instead, began to distribute televisions made in Japan. During this period of marketing a high-tech product made by someone else, the technical community at General Electric had no interest or incentive to keep abreast of the technological advances made in the manufacturing of televisions. At the same time, the foreign producers of the "GE" television had every reason to invest in research and development. This investment enabled them to become more and more competitive. Some analysts now predict that the gap in technology between the manufacturer and the distributor has become so wide that General Electric would have difficulty competing in the market, if the decision was ever made to begin manufacturing televisions again. In fact, the GE Consumer Electronics Group decided in December of 1986 not to renew its contract with Matsushita because of the higher prices demanded by the giant Japanese manufacturer. GE began to produce comparable televisions at its Bloomington, Indiana, plant acquired through a merger with RCA.

It is understandable that when a company or nation gives

up the ability to manufacture, the ability to develop is eventually lost as well. On a national scale, the loss of the ability to manufacture and/or develop could be disastrous.

The United States has enjoyed the status of a world leader for many years, due in large part to the industrial strength of the nation as a whole. This status could very well be jeopardized if our industrial base continues to erode. Modern nations do not prosper or successfully lead unless they have a strong manufacturing base.

MANUFACTURING IN MODERN WORLD HISTORY

Throughout modern history, leading nations have sustained and have been driven by a vital industrial economy.

Consider the history of the European colonial powers from the latter part of the nineteenth century into the first half of the twentieth century: during this period, strong, developed countries invaded and took over the responsibility of governing weak and undeveloped nations.

England colonized India, China, and the Middle East. France colonized Indochina (those countries are called Laos, Cambodia, and Vietnam today). Germany and Italy invaded and colonized parts of Africa. Of course, in the very early years of the colonial period, a host of developed countries colonized the New World, North and South America.

The driving force behind the invasions of the colonial powers was the need for natural resources that did not exist in Europe. These natural resources were required in order to create industries at home and keep them producing. Rubber, lumber, precious metals, iron, tin, bauxite, and, in later years, oil were all necessary ingredients to maintain the vibrant economies of the industrialized nations.

The colonies also provided a natural market for the goods that were manufactured; the colonial powers were able to sell the finished product to the colonies, which did not have the industrial base to process their own natural resources.

Vast wealth was obtained as a result of the colonies' raw materials fueling the industries of the developed nations. With the wealth came power.

England, one of the principal colonizing powers of the eighteenth and nineteenth centuries, was the most powerful nation in the world at that time. Her influence spread around the globe; the pound sterling became the monetary standard of the world, because "the sun never sets on the English Empire." Quite an accomplishment for a country the size of Minnesota.

Prosperity at home accompanied the wealth and power created by a strong industrial base. This prosperity, which overflowed to the general population, improved the standard of living and gave rise to the creation of the educated, working middle class.

Eventually, the colonial period ended. The fates of the former colonies were many and varied as some prospered and some remained underdeveloped.

Once a colony herself, the United States rose to the status of world leader because of natural resources, a strong manufacturing-based economy, and the work ethic of the American people. American technology and manufacturing capability were the marvel and the envy of the entire world. Products made in America were in constant demand even before World War II devastated the industrial base of Europe and Asia.

American influence and the demand for American goods reached an all-time high at the conclusion of World War II. An entire world was rebuilding, and the United States, the only industrial nation with the means to help it rebuild, was feverishly shifting its manufacturing from war materials to consumer goods.

AMERICAN MANUFACTURING TODAY

Products made in America today, however, are not as much in demand as are the products made in Japan. Some products are not even made in America anymore. Like the "General Electric" television, some products are made overseas and marketed by an American company. Considering the influx of foreign products to the United States today, an observer might wonder if the United States of America is slowly becoming a colony once again—an economic colony, but a colony nevertheless.

One can find an historical parallel if one considers American technology to be the natural resource being exported to Asia—American technology that is being utilized overseas to produce high-quality, low-cost products. We purchase these products with dollars that continue to shrink in value when compared to the Japanese yen.

Americans didn't take the first imported Japanese automobiles seriously; we had just grown accustomed to the little Volkswagen "bugs." In the mid 1970s, however, Detroit was taking Japanese imports very seriously.

It wasn't long after the first imported Japanese automobiles arrived that their reputation for high quality was established. Skepticism on the part of the American consumer soon changed to the realization that there was an alternative to Detroit's complacency. Now the prevalence of Japanese cars signals a danger to our nation. Japanese imports, overall, include such a wide variety of products that we, as a nation, should be concerned that we have begun to lose our ability to competitively manufacture.

If industry in the United States does not reverse its trend, we run the risk of eventually losing our status of world leader. Closer to home, the standard of living of every American man, woman, and child could be affected.

HISTORY OF QUALITY IN JAPAN AND THE UNITED STATES

Before we begin to apply solutions to the problem of America's lack of competitiveness, let us first look at how the Japanese made a remarkable transition: in 1945, Japan lay in ashes; today it is one of the world's greatest industrial powers.

World War II: The State of the Economy in Japan and the United States

Japan had entered World War II in order to impose upon its neighbors the "Greater East Asian Co-Prosperity Sphere." This was the scheme whereby Japanese military conquest would provide limitless access to the raw materials of Asia. When the war was over, the Japanese were worse off than ever. Their economy was in ruins and their island nation was as poor in natural resources as ever.

The United States, on the other hand, was enjoying a postwar boom in consumer goods. The giant industrial machine that so greatly contributed to the war effort was now turned loose to satisfy world demand for peacetime products.

Few people today realize that the monumental success of American industry during World War II was due, in large part, to a group of statisticians. These statisticians were responsible for successfully implementing, in American defense industries, control methods that enabled a largely unskilled work force to produce large quantities of high-quality war material.

These innovative efforts were pioneered in the 1930s by Dr. Walter Shewhart of Bell Laboratories.

Two of the statisticians responsible for the refinement and implementation of Dr. Shewhart's methods within the American defense industry were Dr. W. Edwards Deming, and Dr. Joseph M. Juran.

American Quality Pioneers in Japan

In 1950, Dr. Deming was invited to Japan by the Japanese Union of Scientists and Engineers (JUSE); the purpose of his visit was to explain quality control concepts to the Japanese manufacturers. The occupation government under General Douglas MacArthur and leading Japanese industrialists knew that in order to rebuild their economy and gain status as an industrial nation, they would have to change their manufacturing methods, which had caused the phrase "Made in Japan" to become synonymous with junk.

After familiarizing himself with current manufacturing methods in Japan, Dr. Deming called a meeting of forty-five of the leading Japanese industrialists. At this meeting, Dr. Deming informed everyone that if they adopted his method of implementing statistical process control, "they would capture markets within five years the world over; . . . they would take their place beside prosperous nations."[5]

It took a little longer than five years, but no one can deny that by the 1970s the Japanese had performed a minor miracle.

Dr. Joseph M. Juran's contribution to the Japanese industrial turnaround was to teach the Japanese that organizing for continuous improvement was more important than developing technology itself.

Quality Concepts in Postwar America

It is interesting to note that at the same time that the Japanese were being introduced to the concepts of continuous improvement and statistical process control, American industry, which had seen those ideas at work between 1941 and 1945, began to move away from them. The Japanese had nothing to lose by trying these new ideas. American industry apparently felt that there was nothing to be gained by continuing to use proven methods which might slow down production.

In American industry, mass inspection by sample plans allowed production operators to concentrate on quantities of produced items. Complex quality control bureaucracies were established to insure that defective products were not shipped to customers; little attention was paid to solving the problems that caused defective products, however.

Rework of material rejected by quality control became commonplace; some organizations even established rework departments whose sole purpose was to correct mistakes made in production.

The expense of excessive scrap and costly rework operations was addressed by American management by increasing the sale price of the product. Even today, a common practice in American industry is to assign an allowable scrap value to a particular job in order to pass the cost of mistakes on to the customer.

Through the 1960s, the lack of competition caused all of the above practices to become standard in much of American industry. These practices not only caused higher prices, but they also resulted in defective products being sold to the consumer. It is a hard fact that if a poor-quality product is made, all the inspection, sorting, and rework in the world will not prevent some bad products from finding their way to consumers.

Increasing Competition from Japan and the American Response

The Japanese, on the other hand, had been improving upon the basic lessons taught to them by Deming and Juran. By the time the early 1970s rolled around, Japanese imports to the United States were being viewed very favorably by the consumer. In addition to competitive pricing, products made in Japan worked when the consumer took them out of the box or drove them home. Products intended for assembly by the buyer actually had all the parts included. In the rare instance when something was found to be defective, complaints were handled efficiently and in a friendly manner.

World consumers suddenly had an alternative, and they

began to exercise that alternative with a vengeance, causing Japanese-manufactured products to be in ever-increasing demand.

When full realization of the Japanese miracle dawned on American management, some of the responses were less than worthy. Some managers wanted legislation by Congress to protect against imports. The argument was that the low wages paid the average Japanese worker gave Japanese manufacturers an unfair advantage. Certainly, in the 1960s when the dollar was equivalent to 360 yen, the Japanese worker was paid less than his American counterpart, but that didn't explain the glaring difference between the quality of Japanese and American products.

In 1988, the dollar only bought approximately 133 yen. This means that Japanese products, purchased with dollars, are much more expensive today then they were in the 1960s. Yet many people continue to prefer products made in Japan because of their quality.

Many excuses were given for the fact that products made in Japan were of a higher quality than were products made in America. The excuse most often heard was that the Japanese had a different culture that, somehow, made the Japanese worker more aware of the need for quality. Little regard was given to the fact that this was the same culture that, before 1950, produced items that were considered to be very low quality.

The most damaging excuse for the quality differences between American and Japanese products, offered by American management during the 1970s, was directed at the American work force. Many people truly believed that the average American worker no longer took pride in his or her work.

There are many ways to refute the idea that the Japanese worker works harder and better than the American worker because of cultural differences. The best way is by providing examples such as the following.[6]

In 1974, the Quasar Division of Motorola was sold to the

Japanese electronics firm, Matsushita. Prior to 1974, the annual cost due to service calls during warranty periods was on the order of $22 million dollars; the inspection and rework departments employed 120 people. After only four years of Japanese ownership, cost of service calls during the warranty period plummeted to $4 million, an $18 million improvement. Rework also improved to the point where only 15 persons were required for the inspection and rework departments.

These improvements, brought about after the change in ownership, were implemented by the same American work force that had been employed by Motorola. The only real difference was that Matsushita introduced what has popularly become known as **"Japanese-style management"**: a management style the Japanese learned from two Americans—Deming and Juran.

These improvements occurred because the new owners instructed the local work force in the tools and techniques of **statistical process control, continuous improvement,** and the **project team concept** of problem solving.

The American work force at Quasar, second to none, embraced the tools, applied them successfully, improved the competitiveness of Quasar, and insured for themselves continued employment in manufacturing.

It doesn't take a change of ownership to be successful, however. Examples of American-owned companies that have accepted the Deming and Juran philosophies and are actively competing are numerous:[7]

- Defects in Eastman Kodak's Ektaprint copiers are down by 300 percent, production costs down by 70 percent, since 1983. This was accomplished by eliminating product inspectors and by making each worker responsible for his or her own quality.
- Rockwell International sells a key component to Japanese manufacturers of facsimile equipment. The Rockwell modem board enables "fax" equipment to send and

receive digital data via phone lines. Rockwell dominates 70 percent of Japan's market for high-speed modems.

- Armco Inc., a steel producer located in Middletown, Ohio, has improved manufacturing costs by 25 percent, for a savings of $2.5 million per month, since adopting the Deming and Juran philosophies.
- Inland Steel slashed its reject rate of flat-rolled steel in half between 1985 and 1987 as a result of introducing statistical process control.

Briggs and Stratton, Xerox, Harley Davidson, Ford Motor Company, Chrysler, . . . the list goes on and grows longer with each passing day because American management has discovered that their own work force is as competent and as interested in quality as any work force in the world; people need only be given the tools to do it right the first time.

2

The Nature of Variation

The Japanese learned many basic quality concepts from men like Deming and Juran—concepts that, when applied to Japanese manufacturing, helped give birth to the mighty Japanese industrial nation of the 1980s. One of the more basic concepts embraced by the Japanese was the idea that manufacturing processes vary according to natural law; that is, process and product differences that we encounter on a daily basis are the result of the same types of variation that govern the world we live in.

We need not look far to experience **variation;** everything we are associated with is controlled by this natural phenomenon. The weight of newborn babies, the height of high school freshmen, the time it takes us to get to work or return home are all examples of everyday variation.

FORMS OF VARIATION

There are two forms of variation. The first form, **normal variation,** is the type which, by experience, we have come to expect and have learned to live with; the other type, **un-**

13

common variation, comes as a surprise when we encounter it because we are not accustomed to it.

Consider, for example, the everyday experience of walking past strangers on the street. As we walk down the main street of a typical American city we are likely to encounter many individuals of "average" height, some a bit taller and others a bit shorter than average.

Variation in height, within certain limits, is something we take little notice of because we are quite accustomed to the sight of a common range of heights. If, on the other hand, we encounter an individual of uncommon height such as the famous basketball player, Wilt Chamberlain, or Tattoo of the television series "Fantasy Island," our attention is drawn to this uncommon sight, even momentarily, because we have encountered something out of the ordinary.

The common, everyday variation that we take for granted is of a random nature and is totally natural; the other type of variation, the kind we don't often encounter, is out of the ordinary and a result of improbable odds. Common variation is sometimes referred to as natural or normal variation and is presented graphically in Figure 2.1 by a **normal curve.**

The normal curve is characterized by one peak, symmetrically trailing off on both sides to approach the base line.

Any study of natural events finds virtually all of the possible events falling within the normal curve as a result of

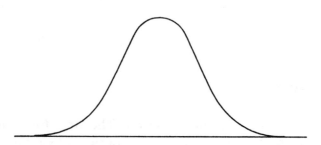

FIGURE 2.1 Normal curve.

natural variation; events falling outside of the curve are the result of unusual circumstances.

This concept of the normal curve can be proven by maintaining an accurate record of the amount of time it takes to complete any repetitive task—for example, recording the amount of time it takes to get to work every day.

Often, when someone is initially questioned regarding the length of time it takes to commute to work on a daily basis, a singular answer such as "twenty minutes," "forty-five minutes," etc. is promptly given. Further probing results in modifiers such as "about twenty minutes," "roughly forty-five minutes," or "approximately a half hour."

Even though everyone is aware that traffic patterns vary—traffic lights are sometimes in their favor and sometimes against them, or natural variations in the weather cause minor delays of public transportation—many people simply do not consider the time it takes to get to work in the context of normal variation.

A study taken over a period of time would probably reveal that, although an individual arrived at work within a predictable time frame the majority of times, an occasional uncommon circumstance, such as road construction, would cause that person to arrive later than normal. On the other hand, the occasional broken alarm clock might cause the same person to break the speed limit and get to work in record time.

HISTOGRAMS

A tool that is used to graphically depict variation is called a **histogram.** The histogram in Figure 2.2 provides a clear and concise picture of the amount of time it took an individual to get to work on a daily basis, over a one-year period. The fifteen-mile distance was covered by automobile, and was a mixture of New England small city traffic conditions, and rural driving.

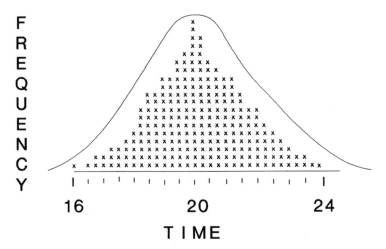

FIGURE 2.2 Histogram depicting normal variation in an individual's travel time to work (employee 1).

The base line or horizontal axis of the histogram represents the time, in minutes, to commute to work each day; the vertical axis represents the number of times in a one-year period the subject, employee 1, arrived within that specified block of time.

Our subject, an exacting creature of habit who has a four-wheel-drive Ford Bronco, leaves the house at approximately 7:30 each morning and, depending on traffic flow, traffic light sequences, the occasional quick stop at the neighborhood mailbox, etc., arrives at work between 16 and 24 minutes later.

The solid line outlining the pattern or distribution in the histogram takes on the shape of the normal curve. This simple study shows that, for employee 1, the variation in the 16-minute to 24-minute process of travelling to work is affected by common, everyday, normal circumstances.

In taking a closer look at the histogram, it also becomes clear that the average travel time for employee 1 is centered around 20 minutes. In other words, the **process aver-**

age of employee 1's travel time is 20 minutes. It also appears that the normal variation in travel time is plus or minus 4 minutes. In other words, the sum total of the normal variation is ± 4.0 minutes. This process, because it falls within the normal curve, is stable, consistent, and predictable. Management and colleagues can rely on this person to arrive at work within a specified time frame because of the consistency due to normal variation. This concept of consistency within a process is very important. When a process is consistent it is said to be **stable**; when it is inconsistent, it is said to be **unstable**. Much can be determined about a process by plotting data in histogram form; for example:

Figure 2.3 provides data on another employee who lives about the same distance from the factory as employee 1. This individual has difficulty waking up in the morning and has an old car that is prone to starting hard, if at all. By outlining the histogram with a solid line, it becomes ob-

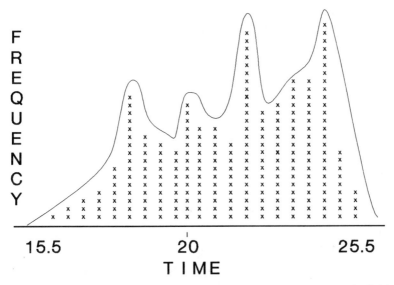

FIGURE 2.3 Histogram depicting uncommon variation in an individual's travel time to work (employee 2).

vious that there is nothing normal about this distribution; no one could predict, on any given day, when this person will arrive to begin work. Employee 2 is inconsistent and unpredictable. In other words, the variation affecting travel time is uncommon and out of the ordinary, and therefore the process of getting to work on time is unstable.

Figure 2.4 represents employee 3, who is predictable with the exception of a few circumstances when abnormal weather conditions causes unforeseen delays. Notice that the process average of 20 minutes for employees 1 and 3 are essentially the same. Employee 3, however, has a number of occurrences outside the curve; the arrival times that fall outside the curve represent the few times, due to unusual weather conditions, that employee 3 was late.

This type of simple data collection exercise can be applied to understanding any process. The horizontal axis could be changed to represent the weight of newborn babies, the height of high school freshmen, or the size of oranges from a particular harvest. As data is collected, the shape of the histogram will remain essentially normal as long as only common cause or normal variation is at work. In cases where

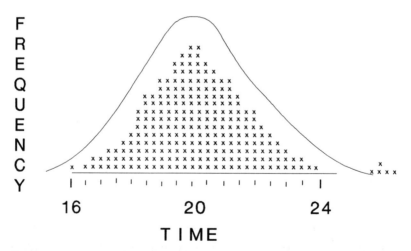

FIGURE 2.4 Histogram depicting occasional uncommon variation in an individual's travel time to work (employee 3).

occasional uncommon variation is at work the majority of data points will fall under the normal curve while a handful of data points will fall outside. Manufacturing processes are influenced by the same common and uncommon variation and can also be represented graphically.

THE ELEMENTS OF A PROCESS

This introduction to process variation would not be complete without further defining the elements that make up a process. After all, the normal variation of an end product is determined by the combination of normal variation found in each of the process elements.

There are four elements in any given **process.** These are shown in Figure 2.5 and include

Equipment
Raw Material
Methods
People

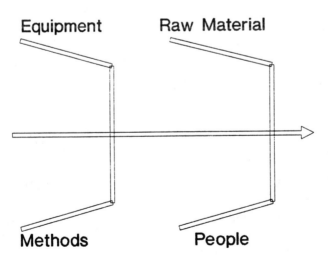

FIGURE 2.5 Four elements that make up any process.

It is interesting to note that customers readily accept the fact that all processes have variation caused by the four process elements. Proof of this statement lies in the accepted practice of providing manufacturers with a minimum and maximum specification for each product dimension or requirement.

Generally speaking, customers provide manufacturers with a **nominal** requirement; this nominal figure is the customer's desired specification. If the customer could remove all variation from a process, and get all product at one dimension, the dimension they would choose would be the nominal dimension. Nominal is what the customer *really* wants. It is only because everyone recognizes that processes have variation that customers include with the nominal requirement a plus and minus range—a **tolerance**.

Equipment

Process variation as a result of equipment stems from the fact that no two pieces of machinery are exactly alike. Also, variation in product happens due to tool wear and fluctuations in equipment speeds, temperatures, vibrations, etc.

Many people in manufacturing don't fully appreciate the fact that the 220- and 440-volt line input, which drives much of industry's equipment, may range as much as ± 10 percent. In other words, the accepted, normal variation of the 440 volts at an extruder may range from 396 volts to 484 volts. With this type of variation, it is reasonable to expect that a production department would experience normal variations in line speed, screw rpm, and zone temperatures. These normal variations will, to some degree, be reflected in product parameter variation such as outer diameter, wall thickness, tensile, and elongation.

Raw Material

Raw material introduces variation into processes because no two batches of raw material are alike. Also there is variation within raw material batches. The melt index identified on a lot of polymer is determined from a sample reading. We cannot expect the melt index of an entire 1,000-pound lot of material to be exactly that of the sample measured by the supplier. Wire and cable manufacturers purchase all raw material according to written specifications which call out desired nominal parameters modified by tolerances.

It is perfectly natural to expect some variation in a final product, such as insulated wire, if the melt index between raw material batches varies within allowable specifications. For example, if a lot of compound with a melt index at the high end of the specification is introduced for one run, we can expect to see differences in the final product if the next batch has a much lower melt index. Also, consider a batch of material that has been identified as having a melt index—based on a measured sample—exactly at nominal. There is every reason to believe that as the material is used, different melt indices will be encountered. Variation between batches, as well as within batches, of raw material causes process variation.

Methods

The process element methods contribute to end product variation in the same manner as equipment and raw material. Drawn, cabled, or extruded wire vary tremendously depending on the method used to produce them. Standard methods or operating procedures are often provided in order to insure a minimum amount of variation in the end product. Unfortunately, even when procedures are provided, operators may change them or ignore them completely; this is discussed in detail further on in this chapter.

People

Variation introduced into a process as a result of people stems from the fact that not everyone does things exactly alike. Even when operators follow standard operating procedures to the letter, differences in setup will be evident. This is particularly true in processes that depend upon operator "feel" for quality product instead of relying on quantitative settings.

Many examples can be found in the wire and cable industry which demonstrate variations due to people. For example, some organizations have standard procedures for setting up extruders in an effort to keep startup scrap to a minimum; some operators follow the procedure to the letter and produce a few pounds of scrap while others, following the same procedure, produce much more scrap. Various levels of operator training will often be reflected in variation of product quality and yield.

Even people who have similar levels of training and experience cause variation to be introduced into a process. This is especially true in operator-dominant portions of the process. Consider the results of two different individuals who have chosen guider tips of slightly different diameters. Each is acting according to his or her experience. Each might make perfectly acceptable product, but the wall thickness variation for one process will probably be totally different from the other.

VARIATION IN MANUFACTURING PROCESSES

Before we can fully understand a process, it is imperative that we accept the concept that manufacturing processes vary in the same manner as everything else around us. This is an important concept that, unfortunately, is not readily accepted by many in industry.

Misunderstanding of Variation—Case 1
(Operator Adjustment at Shift Change)

An example of how the lack of understanding of the concept of normal variation applies to manufacturing processes can be found at shift change in many wire and cable plants. Take notice of the actions of some extruder operators when they come onto a shift and assume responsibility for the quality of the product being produced on their assigned extruder.

Many times the very first task performed by the oncoming operator is one of measurement; this is perfectly normal, because management often has issued the operators some type of measuring device such as an OD tape, micrometer, or vernier caliper. Also, being human, the oncoming operator wants to be certain that he or she has not accepted responsibility for a process which is already making scrap. Operators are often less than totally confident of the ability and attention of the previous operator.

Assume that wall thickness has been identified by management as the product parameter that is to be checked by the operators on a periodic basis in order to insure quality product. The customer has specified that the nominal wall thickness should be .050 and the tolerance should be ± .005. The measurement for wall thickness is generally taken by the operators at every reel change.

For the purpose of this demonstration we will also assume that the process average is centered exactly at nominal and that the entire spread of variation is ± .004; Figure 2.6 represents the process average center and the variation of wall thickness for the process under discussion.

Notice that the process is stable and that, in addition to all product falling within the normal distribution, everything falls within the customer's specifications. This process is **capable** as well as stable. When a process is capable it is providing all product within customer specification. That is, with only normal variation at work, and the process av-

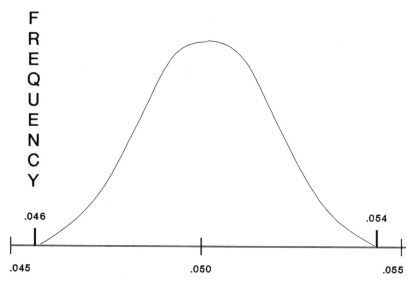

FIGURE 2.6 Normal curve depicting process average of .050 and variation of ± .004.

erage at nominal, the normal variation encountered allows all products to meet specification.

Although our process is predictable in that we can anticipate the maximum wall thickness and the minimum wall thickness to be achieved with this combination of equipment, raw materials, methods, and people, we cannot predict from what portion of the curve a random sample will be chosen. Nor can we determine where the process average is centered based on simple measurement of one or two samples. If our newly arrived operator is unlucky enough to sample the process at the time when normal variation is causing higher walled product to be produced, the operator may draw the conclusion that the process is "running on the high side," and adjust to correct.

The initial sample measured by our oncoming operator is represented by the X on Figure 2.7.

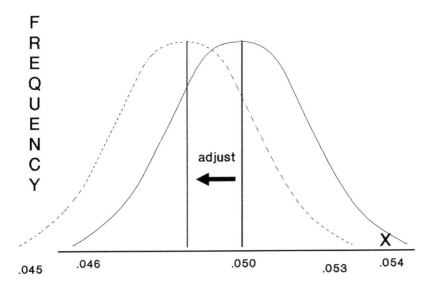

F
R
E
Q
U
E
N
C
Y

adjust

.045   .046   .050   .053   .054

WALL THICKNESS VARIATION

FIGURE 2.7 Effects of operator adjustment to a process.

The effect of the operator adjustment is represented by the dashed curve on Figure 2.7.

With the best of intentions, our extruder operator has turned a process from one producing 100 percent of the product to customer specifications, to one now producing a portion of the product outside the lower limit.

Many operators have a mental image of the process output as being devoid of variation. Operators often think that with a certain combination of dial settings and raw material the output is constantly at a single, discrete point on a continuous scale. Thinking of the process yielding product at one point leads people to believe that changing a dial setting shifts the output to a higher or lower point on the scale. Figure 2.8 represents this imaginary process yield.

It must be clearly understood that manufacturing processes are subject to the same variation that we experience

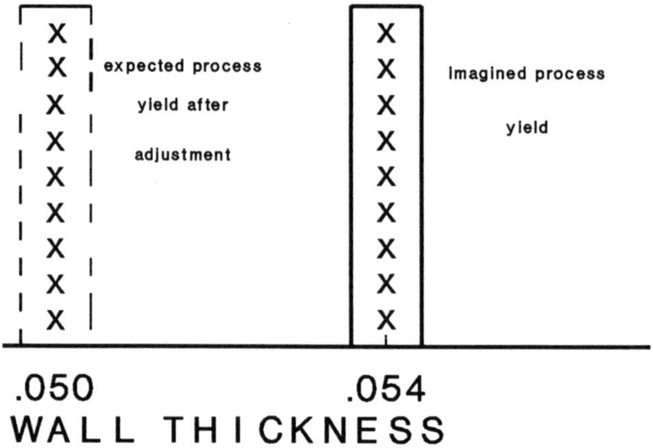

FIGURE 2.8 Process yield without variation.

in nature; the same variation previously discussed in this chapter.

Operator adjustments have the effect of shifting the entire distribution of variation by changing the process center to a point higher or lower on the scale. Because operators have this misconception, anytime a measurement yields a reading which is too near the upper or lower customer specification, an adjustment is made. The end result is depicted in Figure 2.9.

Obviously, the constant shifting of process average causes a great deal more variation in the final product than is necessary or desirable.

It is also important to note that constant adjustments can be counterproductive even if the amount of variation due to common cause is narrow and adjustments do not result in product outside customer specifications. This is because the adjustments cause an increase in the amount of product that is not on or near nominal.

It must be remembered that customers offer tolerances because they have an appreciation that processes vary; everyone knows it is impossible to purchase product with-

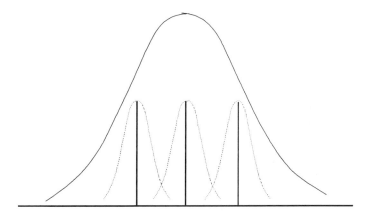

FIGURE 2.9 Undesirable variation due to operator adjustments.

out any variation. Also, it is safe to assume that, in some cases, the nominal requirement is the most desirable; if the customer could purchase all of the product at one point on the continuous scale of measurement they would most likely choose the nominal.

Some Japanese companies have a philosophy that is aimed at understanding the amount of variation present in their processes, and at working to continuously reduce that variation. This constant reduction of variation is an effort to make as much of the product as possible at or very near the nominal, thus improving customer satisfaction.

Operator adjustments work against this concept of providing as much of the product as possible at the point that the customer would truly like it.

Misunderstanding of Variation—Case 2
(Scrap Reduction versus Raw Material Costs)

Producing less scrap is another reason to reduce variation. Process average shifts can be caused either by operator adjustments or by some uncommon variation, such as raw material that is off specification or a voltage reduction beyond what is normally experienced. A process average shift caused by uncommon variation might be absorbed without producing any scrap if the variation was narrow and centered at nominal to begin with.

In the wire and cable industry, a full understanding of variation, how to reduce variation, and how to place the process average center where we want it is imperative to reducing raw material usage.

Consider the example in which a customer has ordered wire to a wall thickness specification of .065 ± .005; a study shows the process to be stable and capable of maintaining a wall thickness of ± .001, but the operators have been purposely running the process towards the high side in order to insure that no product is scrapped due to low wall.

The solid line curve on Figure 2.10 represents where the process has been operating with regard to wall thickness; the broken line curve represents how the process could run if the proper understanding and tools were in place. Obviously, if the process could be run toward the low side, the material savings would surely be a significant contribution to the organization's competitiveness. Additionally, the operators, with their good intentions of insuring that scrap costs be kept to a minimum, were probably using up this imaginary savings in material costs. Even the best intentions can lead a company astray when a proper understanding of variation and the control of variation is not in place.

Variation and Short Production Runs

An understanding of variation can also help the wire and cable industry be more competitive in the area of short production runs. The same misunderstandings that cause an operator to adjust the process based on a single measurement also give rise to excessive startup scrap.

Chapter 6 explains a technique whereby an operator may

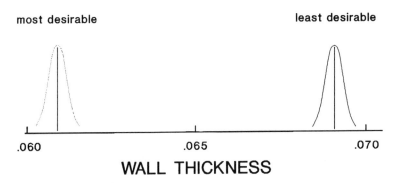

WALL THICKNESS

FIGURE 2.10 Material optimization.

better understand where the process is centered during startup, and thereby reduce the number of startup adjustments and associated scrap.

Variation and Management/Work Force Responsibility

An understanding of processes and process variation even helps define the roles and responsibilities that both management and the work force must assume with regard to the overall process. To understand this point, it is important to review the definitions of process stability and process capability.

Process Stability—A *stable* process yields a consistent output with normal variation.

Process Capability—A *capable* process yields 100 percent of the product to customer specification.

Stability is a function of normal variation; as long as only normal variation is at work, stability is present. The operators, supervisors, setup people, etc. are in the best position to identify periods of instability, and in most cases, to correct for it.

Maintaining stability is the responsibility of the local work force.

Only a local work force trained in the nature of variation and having the tools to measure variation can properly identify stability or a lack of stability. Training and the provision of tools such as control charts, necessary to the identification of stability, are the responsibility of management.

Capability is a function of the four elements of the process: equipment, raw materials, methods, and people. These elements can be chosen only by management. Operators cannot purchase new equipment or choose a different raw material supplier in order to reduce the amount of process variation, making an incapable process capable. In this context, capability is the sole responsibility of management.

A CLEAR PATH TO BECOMING COMPETITIVE

The course is clear for any wire and cable company that truly wants to be competitive. Specifically:

Management must make a commitment to understanding and reducing all process variation as much as possible.

- Management must make a commitment to provide only capable processes to the local work force, and to provide them with the tools and training to recognize when stability does not exist.
- Management must allow the local work force the time to understand process variation and make corrections, when necessary, in order to restore stability—not to just make adjustments to compensate for the lack of stability.
- The local work force must accept the responsibility for learning to use the tools that are provided by management, and conscientiously apply those tools to maintain stability.

- The local work force must be willing to accept full responsibility for the quality of the product; the days of relying upon a roving inspector are over.

A concerted effort by management and the local work force to fully understand and reduce variation, and to use statistical process control to learn as much about the process as possible, can make American wire and cable companies as competitive as their Japanese counterparts.

3

An Introduction to SPC

In the mid-1970s, due to an increasing loss of market share to Japanese manufacturers, the American automotive giants began to appreciate how serious the threat from offshore competition had become. Soon, other American industries began to feel the pinch of their own foreign competitors. Office equipment, consumer electronics, home appliances, and the farm implement market were among the many industries that suddenly realized "business as usual" was no longer satisfactory—if they wanted to survive.

American business people returning from visits to Japan during this period offered many reasons for the Japanese productivity and quality marvel; some of the reasons gave rise to interesting myths which were touched upon in Chapter 1. Some said Japanese success was due to the fact that the average Japanese worker had a better attitude toward his/her job and employer. It was reported that many Japanese workers *ran* from one work station to another instead of walking, like American workers. The formation of a lifetime bond between management and employee was another reason given for the high quality of Japanese products. Of course, many fingers were pointed toward the unique relationship that existed between some Japanese unions and their parent companies.

Among the many subjective excuses presented by some American managers during the 1970s was the real reason for the revitalization of Japan's industry. Simply put, over a period of twenty years, Japanese managers had educated their employees in the tools and techniques of **statistical process control (SPC)**. Japanese managers had fully embraced the philosophies and teachings of men like W. Edwards Deming and Joseph M. Juran and, in so doing, had resurrected a shattered economy and had made the standard of living of the Japanese people the envy of the Far East.

While some American managers were demanding trade protections from Congress, other more enlightened individuals recognized the value of what the Japanese had accomplished and started to learn more about statistical process control themselves. Extensive training programs were instituted in some of the larger American manufacturing organizations.

Statistical process control became the newest American manufacturing buzzword; seminars on the topic became as common as the cold. Unfortunately, many of the available seminars and presentations did not place statistical process control in the context of **continuous improvement**. Nor did they address the matter of statistical process control tools that would apply to wire and cable manufacturers and other industries engaged in the manufacture of continuous or homogenous product.

A FALSE START WITH X BAR AND R CHARTS

The experiences of Typical Wire and Cable Corp., a fictitious company, outlined in the following chapters are intended to illustrate some of the pitfalls that can be encountered by wire and cable companies attempting to launch a statistical process control effort.

Typical Wire and Cable Corp. (TW&C) employs approximately 500 people engaged in the manufacture of specialty wire; TW&C extrudes, draws, braids, bunches, twins, and cables a full range of products to meet the needs of its customer base.

Joe, the quality assurance manager at TW&C, had become aware of statistical process control through several articles he had read in different trade journals. Joe felt sure that TW&C would benefit if he could learn enough about statistical process control so that he could implement a program throughout the plant.

Fortunately, there was enough money in the training budget to allow Joe to attend a three-day statistical process control seminar being presented in a neighboring state.

At the end of three days of intensive training and workshops, Joe returned eager to apply his newfound knowledge to the manufacture of wire and cable. Joe's aim was nothing less than making TW&C the most competitive wire and cable company in the world.

At a presentation to the company president and Joe's fellow department heads, Joe made a strong pitch for an all-out internal training program. Joe explained that this was necessary in order to get everyone up to speed on statistical process control, and to maximize the benefits that he was sure the company would derive from SPC. The president, being a prudent individual, suggested that, before the company would launch into a crash program, Joe should prove the value of his newly acquired skills on a product known to run well. The idea of demonstrating control on a product that had traditionally been a money-maker seemed reasonable to the staff and was accepted by Joe.

To introduce SPC to his company, Joe decided to work with the **X bar and R chart**, also known as the $\overline{X}$-R chart or the average and range chart. Joe had learned at the seminar that this chart was one of the most widely used and best known **control charts**; it was a key tool used in statistical process control. Joe figured that this chart would certainly prove the value of his newly acquired knowledge and would also help convince the president that an all-out internal training program in SPC would pave the way to a bright future for TW&C.

Joe decided to perform his study on the outer diameter of an insulated primary, which he knew was favored by the operators as being a trouble-free job; he also knew that the customer considered the outer diameter to be a critical dimension.

The customer specification was .035 ± .003.

Step One: Collecting Samples and Measurements

Joe had been concerned about the direct application of the X bar and R chart at TW&C because, at the seminar, all the workshops and examples given dealt with the need to take samples and measurements in groups of five or more. Joe knew that although you could select samples of five from the output of a screw machine, you really couldn't stop an extruder, cut out five small samples for measurement, and start up the extruder again.

He solved his dilemma by deciding to take five samples for measurement from the end of each reel; this would work out quite well, because this particular customer required the final product to be put up on 2,000-foot reels, and the next scheduled run was for a minimum of 30,000 feet. A sampling of fifteen subgroups of five each was shy of the minimum he was instructed to use at the seminar, but good enough for demonstration purposes. Also, the operators were in the habit of taking measurements at reel change.

The operators had been asked to measure the diameter

using their hand-held micrometers, and to carry the readings out to the fourth place.

Table 3.1 represents Joe's raw data for five samples from each of fifteen reels.

Step Two: Average and Range Calculations

Joe collected the raw data and, as he had been shown at the seminar, calculated an **average** (X bar or $\overline{X}$) and a **range** (R) for each subgroup of five measurements. The average of any subgroup is calculated by means of summing all the data points in a subgroup and dividing by the number of data points in the subgroup. For example, the average of subgroup 1 is calculated in the following manner:

TABLE 3.1 Sample Measurements of Outer Diameter of Insulated Primary

Subgroup:	1	2	3	4	5
	.0350	.0354	.0350	.0351	.0349
	.0349	.0353	.0350	.0352	.0350
	.0349	.0353	.0351	.0352	.0350
	.0350	.0354	.0350	.0352	.0349
	.0349	.0353	.0350	.0351	.0349
Subgroup:	6	7	8	9	10
	.0350	.0353	.0349	.0350	.0351
	.0348	.0352	.0350	.0349	.0350
	.0350	.0352	.0349	.0350	.0351
	.0350	.0352	.0350	.0349	.0351
	.0350	.0351	.0350	.0349	.0350
Subgroup:	11	12	13	14	15
	.0349	.0352	.0350	.0349	.0352
	.0349	.0352	.0351	.0349	.0350
	.0350	.0350	.0351	.0349	.0351
	.0350	.0351	.0352	.0350	.0350
	.0351	.0353	.0352	.0350	.0352

Subgroup 1 sum = .0350 + .0349 + .0349 + .0350 + .0349 = .1747

Subgroup 1 average = .1747 ÷ 5 (number of data points in
subgroup one) = .03494

The range of a subgroup is determined by subtracting the smallest measurement of a subgroup from the largest measurement in the same subgroup. For example, the range for subgroup 1 is found in the following way:

Subgroup 1 range = .0350 − .0349 = .0001

Joe observed the convention of carrying his average out to one more decimal place than that given in the measured data.

Table 3.2 shows the averages and ranges for each subgroup.

TABLE 3.2 Averages and Ranges of
Outer Diameter Measurements

Subgroup	Average	Range
1	.03494	.0001
2	.03534	.0001
3	.03502	.0001
4	.03516	.0001
5	.03494	.0001
6	.03496	.0002
7	.03520	.0002
8	.03496	.0001
9	.03494	.0001
10	.03506	.0001
11	.03498	.0002
12	.03516	.0003
13	.03512	.0002
14	.03494	.0001
15	.03510	.0002

Step Three: Grand Average and Average Range Calculations

Joe took the exercise to the next step and calculated the **grand average** (X double bar or $\overline{\overline{X}}$) and the **average range** (R bar or $\overline{R}$) by adding up the individual averages and dividing by 15, the number of subgroups, and adding up the individual ranges and dividing by 15.

Grand Average Calculation

$\overline{X}_{total} = .03494 + .03534 + .03502 + .03516 + .03494 + .03496$
$+ .03520 + .03496 + .03494 + .03506 + .03498 + .03516$
$+ .03512 + .03494 + .03510 = \textbf{.52582}$

$\overline{\overline{X}} = \overline{X}_{total}$ divided by 15 (number of subgroups)
$\overline{\overline{X}} = .52582 \div 15$
$\overline{\overline{X}} = \textbf{.03505}$

Average Range Calculation

$R_{total} = .0001 + .0001 + .0001 + .0001 + .0001$
$+ .0002 + .0002 + .0001 + .0001 + .0001$
$+ .0002 + .0003 + .0002 + .0001 + .0002$
$= \textbf{.0022}$

$\overline{R} = R_{total}$ divided by 15 (number of subgroups)
$\overline{R} = .0022 \div 15$
$\overline{R} = \textbf{.00015}$

Step Four: Upper and Lower Control Limit Calculations

Now Joe plugged the grand average and the average range into the formula for $\overline{X}$-R chart **control limits** he had learned at the seminar:

Upper Control Limit (UCL) for $\overline{X} = \overline{\overline{X}} + (A_2 \times \overline{R})$

Lower Control Limit (LCL) for $\overline{X} = \overline{\overline{X}} - (A_2 \times \overline{R})$

Upper Control Limit (UCL) for $R = \overline{R} \times D_4$

Lower Control Limit (LCL) for $R = \overline{R} \times D_3$

The **upper and lower control limits** for the $\overline{X}$-R charts are statistically valid values placed above and below the grand average. Data points plotted against the control limits help us determine if the process is stable or unstable. (The rules to determine stability will be discussed later in this chapter.) A_2, D_3, and D_4 represent constants developed by Dr. Walter Shewhart in the 1930s; these constants are used in conjunction with average range to determine control limits.

The values for A_2, D_4, and other control chart factors can be found in Table 3.3.

The following is the simple arithmetic for the calculations.

$$\overline{X} \text{ UCL} = .03505 + (.577 \times .00015) = .03514$$

$$\overline{X} \text{ LCL} = .03505 - (.577 \times .00015) = .03496$$

$$R \text{ UCL} = .00015 \times 2.114 = .00032$$

$$R \text{ LCL} = .00015 \times 0 = 0$$

TABLE 3.3 Control Chart Factors

Number of Samples in Subgroup (n)	A_2	D_3	D_4	d_2
2	1.880	0.0	3.267	1.128
3	1.023	0.0	2.574	1.693
4	.729	0.0	2.282	2.059
5	.577	0.0	2.114	2.326

Step Five: Plotting the Data

After Joe laid out the control limits for both the X bar and R charts, he plotted the individual averages and ranges against the limits.

Figures 3.1a and 3.1b represent the two charts and the plotted values.

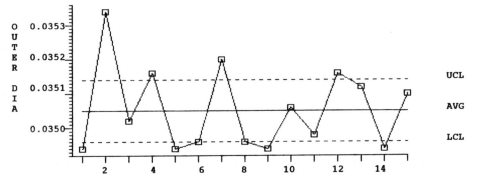

FIGURE 3.1a X bar chart for outer diameter measurements.

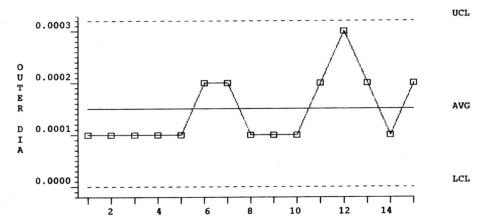

FIGURE 3.1b Range chart for outer diameter measurements.

Step Six: Interpreting the Data

The rules Joe had learned at the seminar for interpreting the control charts were simple and easy to follow. The purpose of the four rules is to help us understand if the process under study is stable and in control.

The best way to understand how the rules apply is to think of the control chart as the **normal distribution** turned on its side with the upper and lower control limits representing the extremes of the distribution.

It's obvious that the two distributions shown in Figure 3.2 are different, and it is necessary to have a conventional way to describe the difference. Statisticians have developed a simple means to communicate differences in width between various distributions. An understanding of how a distribution's width is measured is basic to understanding the rules under discussion.

A convention was developed whereby every normal distribution is divided into six equal segments, three on either side of the centerline. Each segment was termed a **standard deviation** and is designated by the Greek letter "sigma."

Every normal distribution is exactly six standard deviations (or six sigma) in width. If the standard deviation of the narrow distribution in Figure 3.2 is 2.0 units, then the

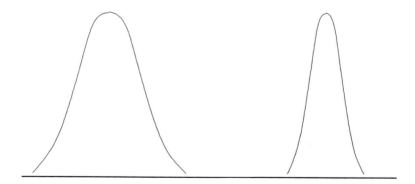

FIGURE 3.2 Normal distributions.

full width of the distribution is 12.0 units. If the wider of the two distributions is twice the width of the other, the standard deviation of the wider distribution would be 4.0 units, and the full width would be 24.0 units.

In viewing a control chart we can picture the upper and lower control limits as representing the tails of the distribution. The distance between the centerline, or average, and the upper or lower limit can be divided into three equal parts representing one, two, and three sigma.

Now for the four rules. (When using subgroups of size four or five the rules apply to both the X bar and R charts.) There is reason to believe instability is present in a process when:

1. One value falls outside of the upper or lower control limit; these limits correspond to the ± three-sigma limits.
2. Two of three consecutive values fall outside the same two-sigma limit.
3. Four of five consecutive values fall outside the same one-sigma limit.
4. Eight consecutive values fall on one side of the centerline.

In practice, only the upper and lower control limits appear on charts (the ± three-sigma limits). The chart would become too busy if all six zones were identified. The one- and two-sigma limits can be drawn in if investigation becomes necessary.

When Joe applied these rules to his control charts (Figures 1a and 1b), he was baffled! The X bar chart showed that eight of the fifteen averages were either above the upper control limit or below the lower control limit. According to the rules, the $\overline{X}$ chart showed this process to be totally out of control. But how could this be? This was a job that never gave anybody trouble, one that was a good money-maker, and no one could remember ever having had a customer complaint related to the diameter of this product.

Joe checked and rechecked his calculations but could find nothing wrong; the sad fact was that one of the best-running processes in the factory did not appear to be in a state of statistical control.

Step Seven: The Conclusion

Joe reported to the president and the staff that although statistical process control would work very well at the Everyday Die Stamping Corp. down the road, or the Super Duper Spring Manufacturing Co. across town, SPC was not for the wire and cable industry.

Thus ended TW&C's first short flirtation with statistical process control.

A Hard Lesson to Be Learned

As a result of a basic difference between two very distinct manufacturing sectors this tale has been repeated across the United States. Manufacturing can be separated into two very general categories: the manufacture of individual, discrete items such as nuts, bolts, or washers, and the manufacture of continuous, batched product such as paper, wire and cable, etc. Companies engaged in the manufacture of wire and cable or chemicals or printed material or packaging material or any process that deals with a continuous or homogenous entity, must learn statistical process control tools other than the traditional X bar and R charts. Joe's basic mistake was in attempting to use the wrong tool.

In Joe's defense, not very many early seminars dealt with any tools other than the garden variety X bar and R chart. When the only tool in your toolbox is a hammer, everything in the world looks like a nail. Joe merely applied the only tool he had available.

WHY THE X BAR CHART DOESN'T ALWAYS WORK

In order to understand why Joe used the wrong tool we must return to the arithmetic by which Joe derived his X bar chart upper and lower control limits.

$$\overline{X} \ UCL = \overline{\overline{X}} + (A_2 \times \overline{R})$$

and

$$\overline{X} \ LCL = \overline{\overline{X}} - (A_2 \times \overline{R})$$

The upper and lower control limits will be an equal distance from the grand average (X double bar) because we add or subtract the same quantity from X double bar; the quantity that we add or subtract is:

$$A_2 \times \overline{R}$$

A_2 is a constant found in Table 3.3; whenever a sample size of five is used, A_2 will be .577. Therefore the dominant factor involved in deciding the control limits is range; as range increases so too will the width of the control limits.

Now consider what Joe was measuring in order to demonstrate that the process was stable/in control. Joe was taking five measurements at reel changeover, five measurements of diameter taken over a relatively short portion of the entire 2,000-foot length. As a practical matter we would not expect to experience very much diameter variation (range) during the short manufacturing time represented by the length of wire being measured in five spots.

A brief review of the data in Table 3.2 shows that there is a maximum .0003 range in any one of the fifteen subgroups. The very small ranges within the subgroups yield a very small average range of .00015; the very small average range results in narrow control limits and, when measurements fall outside these limits, indications of instability.

The X bar chart is best applied to processes in which it is necessary to control variation *within* subgroups as well as variation *between* subgroups. In many continuous processes, such as wire and cable, we would not expect to see appreciable variation within subgroups; in these processes the X bar chart cannot be effectively used.

The reason Joe's control chart showed the process to be out of control is because the X bar chart limits are determined by the variation within each subgroup, which was very small.

THE INDIVIDUAL CHART—A WIRE AND CABLE TOOL

The statistical process control tool that Joe should have applied to TW&C's extrusion process does not look at the range within a discrete distance or reel, but rather looks at the range between reels.

Consider the fact that in an extrusion process we are more concerned with the variation over a long period of time as opposed to a short period; this is because we would expect to see more variation between reels than within reels.

If we refer back to the raw data in Table 3.1 we see that, in fact, we do have more variation between reels than within the distance measured on each reel.

The tool that best addresses variation between batches or subgroups is referred to as the **individual control chart** because, as we will see, the control limits are determined by individual readings over a long period of time (between reels) as opposed to subgrouped readings taken during a short period of time (within reels).

In order to calculate the control limits for the individual control chart we need to determine a range between individual readings. For demonstration purposes, the first reading of each of the original subgroups has been chosen. In practice, it would be necessary to take only one reading from the end of each reel to accomplish our goal. Referring back to the raw data in Table 3.1 and taking the first reading from each reel, we have fifteen individual readings. See Table 3.4.

Adding up all readings and dividing by the number of readings gives us the average.

$\overline{X} = X_{total}$ divided by 15 (number of individual readings)
$\overline{X} = .5259 \div 15$
$\overline{X} = .03506$

Notice how close this average is to the grand average calculated for the X bar and R charts (.03505). The process average is calculated to be almost identical in both cases.

Now we take the range between each reading. (In the equations that follow, the measurements are presented in the order of their occurrence. The fact that the smaller number falls before the larger in some cases is of no consequence. The measurements are simply added algebraically and the polarity—positive or negative—is ignored.)

TABLE 3.4 Individual Readings

Subgroup	Reading
1	.0350
2	.0354
3	.0350
4	.0351
5	.0349
6	.0350
7	.0353
8	.0349
9	.0350
10	.0351
11	.0349
12	.0352
13	.0350
14	.0349
15	.0352

$$.0350 - .0354 = .0004$$
$$.0354 - .0350 = .0004$$
$$.0350 - .0351 = .0001$$
$$.0351 - .0349 = .0002$$
$$.0349 - .0350 = .0001$$
$$.0350 - .0353 = .0003$$
$$.0353 - .0349 = .0004$$
$$.0349 - .0350 = .0001$$
$$.0350 - .0351 = .0001$$
$$.0351 - .0349 = .0002$$
$$.0349 - .0352 = .0003$$
$$.0352 - .0350 = .0002$$
$$.0350 - .0349 = .0001$$
$$.0349 - .0352 = .0003$$

Adding up all the ranges and dividing by 14, the number of ranges, we get R bar (average range).

$\overline{R} = R_{total}$ divided by 14 (number of range readings)
$\overline{R} = .0032 \div 14$
$\mathbf{\overline{R} = .00023}$

Now to get the control limits for the between-reel variation we use a different formula than the one we used for the $\overline{X}$-R chart:

$$UCL = \overline{X} + [3 \times (\overline{R} \div d_2)]$$
$$LCL = \overline{X} - [3 \times (\overline{R} \div d_2)]$$

The value d_2 is another constant that helps us understand where to place the upper and lower control limits about the centerline or process average.

Remember that the upper and lower control limits represent the ± three-sigma limits of a distribution.

For individual control charts, as in X bar and R charts, range is also used to calculate the control limits. Because we have developed each range reading from two individual readings, we choose our constant d_2 based on a subgroup size of two. The value for d_2 can be found in Table 3.3.

Dividing the average range between reels by the constant d_2 gives us one sigma for the distribution of the fifteen individual readings used in this exercise. In order to determine the upper and lower control limits of the distribution we must multiply our answer by three.

It is not recommended to use a range chart with this method because it can be very misleading. The ranges result from the distance between any two individual values, and one extreme individual value results in two extreme ranges because of the more normal individual values on either side of the extreme. This quality of correlated ranges makes it very difficult to identify trends for an individual range chart.

The arithmetic required to calculate the control limits for the raw data collected by Joe looks like the following:

$$\overline{X} \text{ UCL} = .03506 + [3 \times (.00023 \div 1.128)]$$
$$= .03506 + .00061 = .03567$$

$$\overline{X} \text{ LCL} = .03506 - [3 \times (.00023 \div 1.128)]$$
$$= .03506 - .00061 = .03445$$

After the control limits are set about the grand average the individual readings are plotted.

Figure 3.3 illustrates the individual control chart for the raw data taken by Joe. It presents a very different indication of stability compared to what the $\overline{X}$ and R charts were telling us.

SUMMARY

The data used to calculate the individual control chart limits came from the first reading of each original subgroup; any series of readings from these subgroups could have been used. As long as one reading is chosen from each subgroup, the limits will be essentially the same as those calculated.

The individual control chart can be applied in any case that:

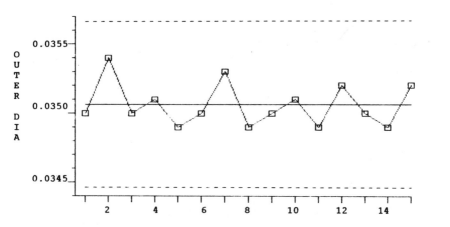

FIGURE 3.3 Individual chart for outer diameter measurements.

- We expect to see more variation between batches than within batches.
- The variation between batches is more important than the variation within batches.
- Each reading represents a unique batch.

The limits of the individual control chart, in essence, represent the ± three-sigma limits of the process variation. For this reason, the limits of this and all processes are sometimes referred to as the **natural process limits.**

In order to determine whether or not a process is stable it is necessary to consider the distance between the limits as divided into their six-sigma increments, and apply the three rules listed below.

When using individual control charts, instability is detected when:

1. One value falls outside the upper or lower control limit; these limits correspond to the ± three-sigma limits.
2. Two of three consecutive values fall outside the same two-sigma limit.

3. Four of five consecutive values fall outside the same one-sigma limit.

Chapter 4 deals further with the individual control chart and its many applications within the wire and cable industry.

4

The Individual Control Chart

Fortunately for TW&C, shortly after Joe decided to hang up statistical process control he came across an article in the *Wire Journal* that discussed individual control charts. After reading the article, Joe realized why his first wire and cable X bar and R chart gave him indications of instability. His first inclination was to address SPC with the staff once again, but he considered that it might be prudent, in the face of the last trial, to bring them something more concrete than theory this time.

After studying several examples of individual control chart applications, Joe decided that he would attempt to demonstrate to the president and staff the usefulness of this tool, using already collected data.

SPC FOR INCOMING INSPECTION

For years the incoming inspection department had been collecting data on many different materials in order to decide whether incoming material was to be accepted or rejected. Although the sample plans used by the inspectors

53

were adhered to religiously, material that did not meet specification sometimes found its way into production and caused significant problems.

A situation such as the one described above existed in the bunching department. For unexplained reasons, wire on some spools would begin breaking while wire on spools from the very same accepted lot would run without any problems.

Step One: Collecting Samples and Measurements

Joe began pouring through the incoming inspection records in order to see if he could apply the individual control chart tool to past information. He wanted to determine if his three suppliers were having periods of instability that were slipping past his inspectors.

Standard operating procedures required his inspectors to take a random sample of thirty spools from a shipment and perform a tensile test on a six-inch piece from each sample spool. The specification called for a minimum tensile strength of .215 pounds on the 44 AWG wire that was experiencing the problems.

Table 4.1 shows Instron readings from the samples tested in the three lots, or shipments, most recently received from each of the three active suppliers.

Joe decided to convert this already compiled information into individual control chart form in order to see if he could detect any abnormalities. The simple arithmetic that follows represents Joe's calculations for the individual control chart limits.

Step Two: Calculating the Averages

Joe calculated the average for each of the samples taken from the three suppliers.

TABLE 4.1 Sample Measurements of Tensile Strength

Supplier A Sample Measurements (X)	Supplier B Sample Measurements (X)	Supplier C Sample Measurements (X)
.214	.264	.253
.225	.230	.267
.232	.235	.252
.233	.237	.250
.217	.245	.249
.218	.275	.257
.205	.224	.252
.213	.245	.257
.225	.278	.253
.227	.235	.257
.217	.238	.247
.230	.287	.254
.217	.269	.264
.217	.263	.263
.225	.262	.264
.205	.261	.255
.227	.237	.267
.225	.270	.247
.220	.260	.266
.213	.230	.253
.218	.260	.267
.225	.223	.270
.213	.268	.247
.217	.217	.250
.217	.227	.267
.227	.228	.265
.225	.258	.249
.226	.270	.266
.207	.268	.254
.214	.216	.247

Average $\overline{X} = X_{total}$ divided by number of samples

Supplier A: $\overline{X} = 6.594 \div 30 = \mathbf{0.2198}$
Supplier B: $\overline{X} = 7.480 \div 30 = \mathbf{0.2493}$
Supplier C: $\overline{X} = 7.709 \div 30 = \mathbf{0.2570}$

Step Three: Calculating the Ranges and Average Ranges

The average range was the next piece of information Joe needed. Because each sample was considered a separate batch, Joe needed to calculate the range between each sample (see Table 4.2), then calculate the average range by adding the ranges and dividing by the number of ranges calculated (one less than the total number of samples).

TABLE 4.2 Ranges Between Individual Batches

Supplier A	Supplier B	Supplier C
.214 ⎫ .011	.264 ⎫ .034	.253 ⎫ .014
.007 ⎧ .225 ⎭	.005 ⎧ .230 ⎭	.015 ⎧ .267 ⎭
⎩ .232 ⎫ .001	⎩ .235 ⎫ .002	⎩ .252 ⎫ .002
.016 ⎧ .233 ⎭	.008 ⎧ .237 ⎭	.001 ⎧ .250 ⎭
⎩ .217 ⎫ .001	⎩ .245 ⎫ .030	⎩ .249 ⎫ .008
.013 ⎧ .218 ⎭	.051 ⎧ .275 ⎭	.005 ⎧ .257 ⎭
⎩ .205 ⎫ .008	⎩ .224 ⎫ .021	⎩ .252 ⎫ .005
.012 ⎧ .213 ⎭	.033 ⎧ .245 ⎭	.004 ⎧ .257 ⎭
⎩ .225 ⎫ .002	⎩ .278 ⎫ .043	⎩ .253 ⎫ .004
.010 ⎧ .227 ⎭	.003 ⎧ .235 ⎭	.010 ⎧ .257 ⎭
⎩ .217 ⎫ .013	⎩ .238 ⎫ .049	⎩ .247 ⎫ .007
.013 ⎧ .230 ⎭	.018 ⎧ .287 ⎭	.010 ⎧ .254 ⎭
⎩ .217 ⎫ .000	⎩ .269 ⎫ .006	⎩ .264 ⎫ .001
.008 ⎧ .217 ⎭	.001 ⎧ .263 ⎭	.001 ⎧ .263 ⎭
⎩ .225 ⎫ .020	⎩ .262 ⎫ .001	⎩ .264 ⎫ .009
.022 ⎧ .205 ⎭	.024 ⎧ .261 ⎭	.012 ⎧ .255 ⎭
⎩ .227 ⎫ .002	⎩ .237 ⎫ .033	⎩ .267 ⎫ .020
.005 ⎧ .225 ⎭	.010 ⎧ .270 ⎭	.019 ⎧ .247 ⎭
⎩ .220 ⎫ .007	⎩ .260 ⎫ .030	⎩ .266 ⎫ .013
.005 ⎧ .213 ⎭	.030 ⎧ .230 ⎭	.014 ⎧ .253 ⎭
⎩ .218 ⎫ .007	⎩ .260 ⎫ .037	⎩ .267 ⎫ .003
.012 ⎧ .225 ⎭	.045 ⎧ .223 ⎭	.023 ⎧ .270 ⎭
⎩ .213 ⎫ .004	⎩ .268 ⎫ .051	⎩ .247 ⎫ .003
.000 ⎧ .217 ⎭	.010 ⎧ .217 ⎭	.017 ⎧ .250 ⎭
⎩ .217 ⎫ .010	⎩ .227 ⎫ .001	⎩ .267 ⎫ .002
.002 ⎧ .227 ⎭	.030 ⎧ .228 ⎭	.016 ⎧ .265 ⎭
⎩ .225 ⎫ .001	⎩ .258 ⎫ .012	⎩ .249 ⎫ .017
.019 ⎧ .226 ⎭	.002 ⎧ .270 ⎭	.012 ⎧ .266 ⎭
⎩ .207 ⎫ .007	⎩ .268 ⎫ .052	⎩ .254 ⎫ .007
.214 ⎭	.216 ⎭	.247 ⎭

Average Range $\overline{R} = R_{total} \div$ number of range calculations

Supplier A: $\overline{R} = 0.238 \div 29 = \mathbf{0.0082}$
Supplier B: $\overline{R} = 0.672 \div 29 = \mathbf{0.0232}$
Supplier C: $\overline{R} = 0.274 \div 29 = \mathbf{0.0094}$

Step Four: Calculating the Upper and Lower Control Limits

Once Joe had the average and the average range, he could then calculate the upper and lower control limits using the formula for the individual control chart.

$$UCL = \overline{X} + [3 \times (\overline{R} \div d_2)]$$
$$LCL = \overline{X} - [3 \times (\overline{R} \div d_2)]$$

Note that for each control chart the calculated number $[3 \times (\overline{R} \div d_2)]$, also called the natural process limits, is first added to $\overline{X}$ to determine the UCL and then subtracted from $\overline{X}$ to determine the LCL.

Supplier A: $[3 \times (\overline{R} \div d_2)] = 3 \times (0.0082 \div 1.128) = 3 \times 0.0073 = \mathbf{0.0219}$

$UCL = \overline{X} + 0.0219 = 0.2198 + 0.0219 = \mathbf{0.2417}$
$LCL = \overline{X} - 0.0219 = 0.2198 - 0.0219 = \mathbf{0.1979}$

Supplier B: $[3 \times (\overline{R} \div d_2)] = 3 \times (0.0232 \div 1.128) = 3 \times 0.0206 = \mathbf{0.0618}$

$UCL = \overline{X} + 0.0618 = 0.2493 + 0.0618 = \mathbf{0.3111}$
$LCL = \overline{X} - 0.0618 + 0.2493 - 0.0618 = \mathbf{0.1875}$

Supplier C: $[3 \times (\overline{R} \div d_2)] = 3 \times (0.0094 \div 1.128) = 3 \times 0.0083 = \mathbf{0.0249}$

$UCL = \overline{X} + 0.0249 = 0.2570 + 0.0249 = \mathbf{0.2819}$
$LCL = \overline{X} - 0.0249 = 0.2570 - 0.0249 = \mathbf{0.2321}$

Step Five: Plotting the Data

Figures 4.1, 4.2, and 4.3 represent the three individual control charts developed by Joe for the three suppliers.

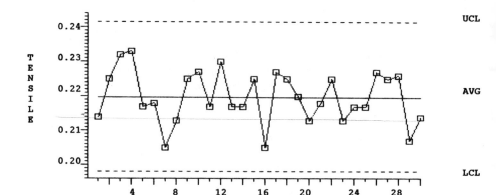

FIGURE 4.1 Individual control chart for Supplier A.

Step Six: Interpreting the Data—Control Charts versus Sample Data

If the decision to accept or reject the lot was based on the sample data alone, Supplier A would be rejected because of the eight values found to be below the .215 minimum. Also, the lots from Suppliers B and C would be accepted because no values fall below minimum. However, if the decision to

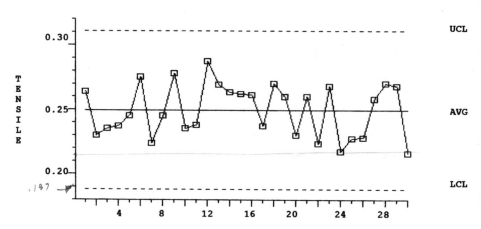

FIGURE 4.2 Individual control chart for Supplier B.

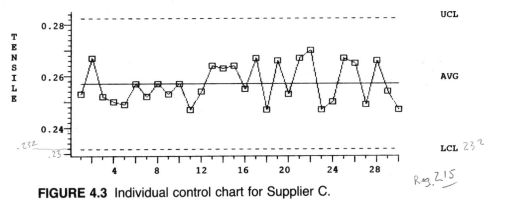

FIGURE 4.3 Individual control chart for Supplier C.

accept or reject is based on the information from the control charts, a different picture begins to emerge.

Certainly lot A is still rejectable material. But lot B is also rejectable, even though no one individual sample falls under the minimum specification. The control chart clearly shows that lot B contains material that will have a tensile strength as low as .188 pounds, since the lower control limit is .1875.

The reason for conductor breaks that seemed to come and go became a little clearer after Joe laid out the three charts. Although the average tensile strength of material supplied by Company B was higher than Company A, the natural process limits calculated in step four show that the amount of variation in the tensile strength was considerably greater in Company B's material than in the material from Company A. For Company B, the natural process limits were .0618 above and below the average, whereas for Company A, the natural process limits were just .0219 above and below the average. This meant that although no individual Company B sample was found to be under specification, one could, while running this lot, expect to see material with a tensile strength as low as .188 pounds. Granted there would not be a very high percentage of material that low, but that would tend to make the solution all the more elusive.

Joe noted with interest that his inspectors, following the rules of the sample plan, would have accepted the lot from Company B; his inspectors would have unwittingly introduced a small percentage of substandard raw material into the process.

Based on the individual control chart information, Joe rejected the material from Supplier A *and* Supplier B. Joe then requested that the purchasing department return the material to Supplier B with a full explanation, since he realized the supplier might have some difficulty in understanding why the material was being rejected.

Joe decided to do a little more analysis work with the charts. He made a comparison between the individual control charts for the material supplied by Companies A and C. The amount of variation evident on each of the two charts was almost identical, but the process average of Company C's material (.2570) was considerably higher than the process average of Company A's material (.2198).

Joe requested that the purchasing department share this information with Supplier A in order to see what could be done to raise their process average on future shipments.

Joe wrote a report that outlined his findings and made strong recommendations that TW&C purchase the majority of 44 AWG conductor from Company C until such time as Company A could raise their process average and/or Company B could reduce the amount of their process variation. Joe formally presented the report at the next monthly staff meeting. He included data that showed a significant decline in the number of wire breaks at the bunching operation since the time that the decision was made to use only Company C material. Joe also took the opportunity to re-present the extrusion information (given in Chapter 3), which he had originally put into X bar and R chart form; this time, however, he presented the data in individual control chart form (as shown at the end of Chapter 3).

Inspectors were instructed to continue taking thirty samples from each incoming lot, but instead of releasing the material based on the individual readings they were instructed to plot their readings on a control chart.

APPLICATION OF THE INDIVIDUAL CONTROL CHART

The control chart that the inspectors used for Supplier C's material was based on Joe's original calculations. Readings from any subsequent incoming lot could also be plotted on the same chart in order to see whether or not the supplier's process had changed in any way.

By using the following criteria for detecting instability, as set forth in Chapter 3, TW&C could insure against accepting any lot that might have any amount of material out of specification. Also, TW&C could tell if there had been a process average shift for any reason.

- One value falls outside the upper or lower control limit (the ± 3-sigma limit).
- Two of three consecutive values fall outside the same 2-sigma limit.
- Four of five consecutive values fall outside the same 1-sigma limit.

Figure 4.4 is a chart clearly indicating a process average shift between lot 178 and lot 179 received at different times.

Both lots were accepted by TW&C, but the process average shift was communicated back to the supplier. It was discovered that lot 179 had been processed through a new annealing oven that had not been properly calibrated.

The application of the individual control chart in this example touches upon a number of interesting statistical process control concepts that are often missed by companies just beginning a statistical process control program.

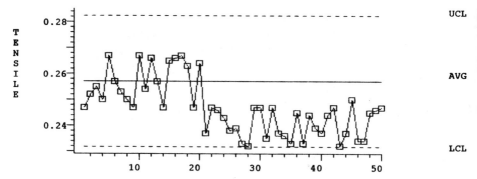

FIGURE 4.4 Individual control chart for Supplier C lots 178 and 179.

Reducing Variation

All too often companies introduce SPC, in the form of control charts, as a means to inform operators when to make adjustments. This is absolutely wrong! The purpose of statistical process control is to understand and reduce variation.

As was discussed in Chapter 2, the only effect of operator adjustments is to change the process average—not to reduce the amount of variation. Control chart indications of instability mean either that something unusual has occurred in one of the four elements of the process or that there has been a process average shift. A process average shift could be due to a change of raw material, the results of a different operator technique, etc.

When Joe used the individual control charts to identify the process averages and the amount of variation in each of the three suppliers' processes, he was making the highest and best use of the tool.

Problem Solving

In our example, there was a chronic problem of excessive wire breaks at the bunching operation. This problem, like many others, had been experienced for years, and operators and supervisors had learned to put up with it. Little data had been collected during the life of the problem; without data it is virtually impossible to solve any problem.

In the mid-1800s the English Lord Kelvin said: "When you can measure what you are speaking about and express it in numbers, you know something about it, but when you cannot express it in numbers your knowledge is . . . unsatisfactory." Nothing has changed in this regard; without a proper analysis of the available data this chronic problem would have continued for more years—years of lowered productivity and competitiveness.

Decision Making

Making business decisions rests squarely with management. All too often, however, management makes decisions without sufficient data.

Buyers make decisions to purchase raw material based on price and delivery alone. Quality professionals make decisions to ship or accept material based on the result of sample plans. Engineers make decisions to install equipment without fully understanding the equipment capabilities. All of these decisions, and others, can be made with a much higher degree of confidence when data that has been properly analyzed is made available.

In our example, management could confidently decide to hold shipment of all material from Supplier B, reject Supplier A material, and purchase future material from Supplier C until conditions at the other vendors improved. These decisions could be made with confidence because the proper statistical process control tool was applied.

Quality and Productivity Improvement

If Joe had not placed the data that had already been collected in individual control chart form, he never would have seen the differences in supplier performance. All too often people in American industries do all the work but fail to receive all the benefit.

In our example, all of the samples had been collected, measured, and recorded. The decision to accept had been made by comparing the readings to a specification; no effort was made to understand the amount of variation, or the risk of accepting poor quality material. With just a little more arithmetic, so much more can be gained in the way of quality, productivity, and operator satisfaction.

PROCESS CONTROL VERSUS PRODUCT CONTROL

As shown in Figure 4.5, a process has four elements: equipment, raw material, methods, and people.

Any time statistical data is collected for the express purpose of understanding and reducing the variation in one of these four elements, we are engaged in statistical process control in its purest form.

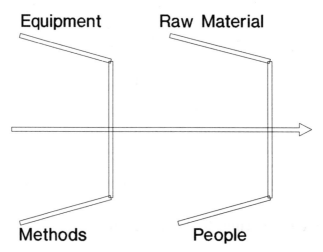

FIGURE 4.5 Four elements that make up any process.

The alternative is to attempt to understand and reduce the variation of the *process* by measuring and analyzing the *product*. Unfortunately, many companies never get beyond the point of measuring and plotting product measurements on control charts.

Joe used Statistical Process Control to understand and reduce the variation of the raw material. In so doing, Joe stumbled upon a concept that many companies never understand; the need to ultimately use SPC *to reduce the variation of the product by reducing the variation within the process.*

More about process versus product control is offered in the next chapter.

5

Process Control versus Product Control

THE ROLE OF CONTROL CHARTS
IN STATISTICAL PROCESS CONTROL

One of the more basic misconceptions in American industry regarding statistical process control is the idea that the presence of control charts equals SPC, and the more control charts the better. In reality, control charts are a means to an end, the end being the overall reduction of process variation.

As was mentioned in the last chapter, the purpose of statistical process control is to help us understand the amount of variation that exists in a process and to show us when to take actions to reduce that variation. Control chart techniques lend themselves very well to the comprehension and reduction of process variation, and when they are successfully used for this purpose they can eventually become obsolete.

Manufacturing organizations that subscribe to the "control charts equal SPC" school of thought are often going through the motions in order to satisfy a customer demand that proof of statistical process control be demonstrated.

In an environment such as the one described above, where the motivating force for SPC is customer requirements, control charts become little more than window dressing. Even in the event that the proper chart technique is being applied, and that the operators have been instructed in the mechanics of taking measurements and plotting data, the organization is engaged in **statistical product control** as opposed to statistical process control.

THE IMPORTANCE OF THE PROPER MEASUREMENT SYSTEM FOR PROCESS CONTROL

A Problem with Monitoring Wall Thickness Using the Individual Control Chart

Consider a situation that Joe, our quality manager at TW&C, came across shortly after he discovered how to apply individual control charts to incoming material.

Working with an eager young production supervisor, Joe was introducing a control chart at extruder number six with the intention of monitoring wall thickness. This particular extruder had been chosen because it was presently running a product for Gigantic Wire Using Corporation. Gigantic's supplier quality engineer was due for a visit and everyone knew that Gigantic was interested in dealing only with wire and cable companies that could demonstrate SPC.

An individual chart was the tool Joe intended to use at extruder number six. Joe had already spent a considerable amount of time training the supervisor and the operators in the mechanics of collecting data in preparation for calculating the control limits. The product characteristic chosen for the exercise was wall thickness. The wall thickness requirement was .030 ± .004. The operators were in the habit of checking wall thickness by taking a sample from the end of each completed reel of wire. In order to make sure both the high and low tolerances—.026 to .034—were being achieved, two dimensions were taken on each sample, a high wall and a low wall reading.

For the purposes of the control chart study, Joe decided to record and plot only the high wall readings. He collected the data from the first eight hours of running time, taking readings at every reel changeover for a total of sixteen measurements. He then calculated the upper and lower control limits and plotted the data according to the instructions outlined in Chapter 3.

Table 5.1 represents the raw data collected by the first shift operator; the readings were taken using the 10-power magnification microscope at the extruder.

Joe made the necessary calculations, as outlined in Chapter 4 (also listed in the Appendix), to determine the upper and lower control limits for the individual control chart, with the following results:

TABLE 5.1 Sample Measurements of High Wall Thickness

Time	High Wall
8:00	.035
8:30	.035
9:00	.035
9:30	.036
10:00	.036
10:30	.036
11:00	.036
11:30	.036
12:00	.036
12:30	.035
1:00	.035
1:30	.035
2:00	.035
2:30	.035
3:00	.036
3:30	.036

$$\overline{X} = .568 \div 16 = .0355$$
$$\overline{R} = .003 \div 15 = .0002$$

$$3(\overline{R} \div d_2) = 3(.0002 \div 1.128) = 3(.00018) = .00054$$

$$UCL = .0355 + .0005 = .0360$$
$$LCL = .0355 - .0005 = .0350$$

Figure 5.1 illustrates the individual control chart developed by Joe from the raw data.

Joe was disturbed! First of all, the pattern of random variation Joe expected to see was not there. This chart didn't look at all like the incoming tensile strength charts (Figures 4.1 through 4.3). Also, the chart indicated that uncontrolled variation was at work; the rule that required that there be no two of three consecutive values outside the same two-sigma limit (see Chapter 3) was being violated. Last, and most distressing, all the product was being run above specification. Even though the high wall thickness probably would not give the customer any cause for complaint, Joe realized that a good deal of expensive raw material was being used unnecessarily.

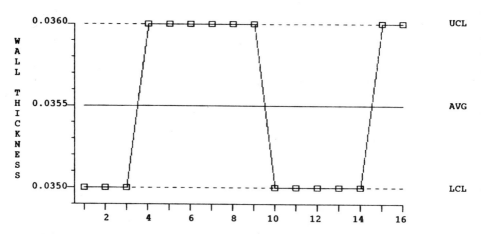

FIGURE 5.1 Individual control chart for wall thickness (measurements taken with a 10 X microscope).

Joe felt like he was right back where he had started with regard to the real world application of SPC.

This is the point in a story where unexpected help arrives just in time to point the hero in the proper direction. In this case, the unexpected help arrived in the person of Mr. John D. Wright, the supplier quality engineeer from Gigantic Wire Using Corp.

The Need to Examine Measurement Systems

John D. Wright (the D. stands for "Do-it") had extensive experience in helping Gigantic's supplier companies implement statistical process control. His philosophy was that SPC had to benefit the supplier in order for the supplier to continue to use SPC after his departure; he was not interested in window dressing.

After spending an hour with Joe, Mr. Wright was very impressed with what Joe had already learned about individual control charts, and his application of this tool to incoming material. A brief review of the latest attempt at application of SPC to the extrusion line caused Mr. Wright to smile and assure Joe that the answer to his problem was simple and easily resolved.

Mr. Wright explained that many organizations attempt to gather data and plot values on control charts without ever attempting to fully understand the measurement system that is being used to collect the data. This lack of understanding of the measurement system will not cause any difficulty if an organization is interested only in impressing people with the presence of control charts. Neither will the need to understand the measurement system be appreciated if a company is engaged in statistical *product* monitoring, using control charts only as a tool to tell their operators when to make adjustments. However, if control charts are applied for the purposes of statistical *process* control, that is, applied for the purposes of understanding process variation and reducing it, then a thorough knowledge of the measurement system is absolutely necessary.

UNDERSTANDING THE MEASUREMENT PROCESS

Measurement systems are indeed processes. Measurement systems have all the four basic elements of any other process: equipment, raw material, methods, and people. In the context of measurement systems, the equipment is the measurement tool and the raw materials are the items being measured.

As a process, a measurement system is subject to the same laws of variation that apply to all other processes. Variation within measurement systems can have a most significant effect on control chart presentations when the systems are not accurate, repeatable, and/or do not offer enough **discrimination**, or the ability to tell one sample from another.

Accuracy

Accuracy is best defined as a comparison to some known standard. Accuracy in the wire and cable industries is most often addressed through a formal gauge calibration program and operator training. Many companies require all measurement tools to be periodically calibrated to standards set by the National Bureau of Standards. A formal calibration program coupled with operator training insures accuracy of measurement tools. For instance, operators who have been properly trained in the use of micrometers make a common practice of checking their already calibrated micrometers against a known standard, such as a gauge block, before measuring product.

Repeatability

Even the most accurate measurement system in the world is subject to some degree of inconsistency (lack of **repeatability**) due to the person or persons using the equipment. Remember, since measurement is a process, we cannot exclude people or methods from the study of any measurement system.

Consider two people measuring the diameter of the same piece of wire, both using the same micrometer. Each individual holds the micrometer differently; each has a different "feel" or sense of when the micrometer anvils have closed on the sample sufficiently well to take a reading. Each person also measures at a different spot on the sample.

The differences discussed above all contribute to some degree of inconsistency. These differences are compounded, of course, if more than one micrometer is introduced, because then we have the potential of slight differences between the tools.

In summary, the lack of acceptable repeatability can be due to different skill levels of operators, different methods applied by equally skilled operators, and/or limitations of the equipment being used. This is demonstrated later in this chapter.

Discrimination

The purpose of statistical process control is to understand and reduce variation. In order to accomplish this goal it is necessary to be able to see the variation by means of measurements. If the measurement system cannot tell the difference between one sample and another, variation will not be detected.

Consider purchasing a gross of inexpensive ball point pens, and setting about to determine how much variation exists in the length of pens. Two implements are available

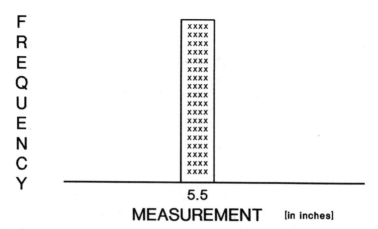

5.5

MEASUREMENT   [in inches]

FIGURE 5.2 Histogram depicting pen length (measurements taken with a ruler). Note—each x represents two measurements.

with which to take data: a standard grade-school ruler and a six-inch vernier caliper. Two histograms should be constructed, one for the measurements taken with the ruler and one for the measurements taken with the vernier.

The smallest unit of measurement on the ruler is 1/16 of an inch, or approximately .060 of an inch. Because the amount of variation between pens is considerably less than 1/16 of an inch, the histogram for the 144 measurements looks like the one illustrated in Figure 5.2.

The smallest unit of measurement on the vernier is .001 of an inch; the vernier, because it has far greater ability to discriminate or tell the difference between samples, gives us a histogram like the one in Figure 5.3. Using the vernier, we can get an appreciation for the amount of variation that exists within the product parameter of pen length.

MEASUREMENT ERROR ANALYSIS

Mr. Wright suggested that a simple **measurement error analysis** be used to determine whether the measurement

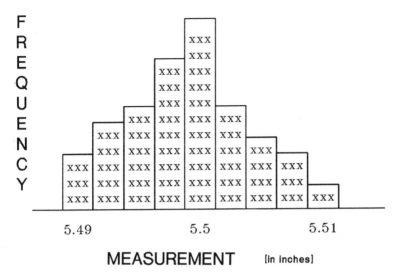

FIGURE 5.3 Histogram depicting pen length (measurements taken with a vernier caliper).

system was sufficient for the task at hand, in order to demonstrate several of the points he was making with regard to measurement processes.

The samples that had been measured for the control chart in Figure 5.1 had been saved; five of these samples were chosen for the measurement error analysis.

Step One: Collecting Samples and Measurements

Joe was asked to have the operator measure all five samples, using the microscope at the extruder, and record the high wall thickness for each of the samples. Next, Joe was asked to have a machinist measure the same five samples using a machinist microscope, and record the high wall thickness for each.

The machinist microscope consisted of a precision slide

that could be moved along the X and Y axis by means of two small hand wheels; the amount the slide moved was displayed on a digital readout. Using this microscope the machinist mounted the sample on the slide and lined up the outside wall with the cross hair. He then zeroed out the digital reading, traversed the slide until the cross hair was lined up with the inside of the wall, and recorded the display on the digital readout.

Two hours later the same operators were asked to measure the same samples again. No mention was made that these were the same samples measured previously; this precaution was taken to insure against operator bias.

The exercise was repeated three more times. Each time the operators measured the samples, they had no knowledge that they were repeating measurements on the same samples. Also, each time the samples were measured they were given to the operators in a different order.

Table 5.2 represents the data collected using the two different microscopes.

Upon reviewing the two sets of data, Joe first noticed that the measurements taken with the machinist microscope were four-decimal-place readings, whereas the measurements taken with the microscope at the extruder were only three-place readings. As in the case of the six-inch ruler and the vernier caliper, the machinist microscope had a smaller unit of measurement than did the extruder microscope, and therefore the machinist's microscope had better discrimination.

Joe realized immediately that the four-place measurements would help him better understand the product variation. When he compared the ranges of the two sets of data, he realized that using the extruder microscope *could* lead to the belief that the production process had far more variation than might actually be the case.

Joe began to get a new appreciation for measurement systems. But Mr. Wright was not through yet.

TABLE 5.2 Repeated Measurements of Minimum Wall Thickness Using Two Different Microscopes

	EXTRUDER MICROSCOPE				
	Subgroup 1 Meas.	Subgroup 2 Meas.	Subgroup 3 Meas.	Subgroup 4 Meas.	Subgroup 5 Meas.
	.034	.035	.034	.035	.035
	.033	.035	.034	.035	.035
	.034	.035	.034	.034	.034
	.033	.033	.035	.035	.035
	.035	.035	.036	.035	.036
$\overline{X}$	.0338	.0346	.0346	.0348	.035
R	.002	.002	.002	.001	.002

	MACHINIST MICROSCOPE				
	Subgroup 1 Meas.	Subgroup 2 Meas.	Subgroup 3 Meas.	Subgroup 4 Meas.	Subgroup 5 Meas.
	.0346	.0352	.0342	.0355	.0352
	.0342	.0348	.0344	.0352	.0355
	.0340	.0350	.0342	.0356	.0352
	.0344	.0345	.0346	.0350	.0352
	.0342	.0352	.0349	.0350	.0355
$\overline{X}$	.03428	.03494	.03446	.03526	.03532
R	.0006	.0007	.0007	.0006	.0003

Step Two: Average, Range, and Control Limit Calculations

Mr. Wright and Joe then sat down to convert the data to control chart form. Because Joe was interested in evaluating the *measurement process* instead of the *production process*, he was instructed to use the X bar and R chart tool for this exercise. Mr. Wright suggested using the X bar and R chart because each series of measurements taken on the same sample, by the same person, at different times represented a subgroup. The average of all the within-subgroup ranges sets the limits for the X bar and the R charts. If a measurement process is repeatable over time, there will not be very much range within the subgroups, and the X bar control limits will be very narrow. If, on the other hand, there is a wide range among the readings within subgroups due to poor measurement repeatability, the control limits will be very wide. In short, the X bar and R chart tool gives us a very graphic presentation of the amount of variation in a measurement system.

Thus Joe was instructed to treat each set of data as a subgroup with a sample size of five. Joe then calculated the average and range for each subgroup, as well as the grand average and average range and the upper and lower control limits for each of the microscopes, using the following formulas for the X bar and R charts. (Formulas and factors are also listed in the Appendix.)

Extruder Microscope

$$\text{Average } (\overline{X}) = X_{total} \div \text{ number of samples}$$

Subgroup 1: $\overline{X} = .169 \div 5 = .0338$
Subgroup 2: $\overline{X} = .173 \div 5 = .0346$
Subgroup 3: $\overline{X} = .173 \div 5 = .0346$
Subgroup 4: $\overline{X} = .174 \div 5 = .0348$
Subgroup 5: $\overline{X} = .175 \div 5 = .0350$

$$\text{Grand Average } (\overline{\overline{X}}) = \overline{X}_{total} \div \text{ number of samples}$$

$$(\overline{\overline{X}}) = .1728 \div 5$$
$$(\overline{\overline{X}}) = .03456$$

$$\text{Range } (R) = \text{Largest measurement in a subgroup} - \text{smallest measurement in the same subgroup}$$

Subgroup 1: $R = .035 - .033 = .002$
Subgroup 2: $R = .035 - .033 = .002$
Subgroup 3: $R = .036 - .034 = .002$
Subgroup 4: $R = .035 - .034 = .001$
Subgroup 5: $R = .036 - .034 = .002$

Average Range $(\overline{R}) = R_{total} \div$ number of samples

$$\overline{R} = .009 \div 5$$
$$\overline{R} = .0018$$

UCL for $\overline{X} = \overline{\overline{X}} + (A_2 \times \overline{R}) = .03456 + (.577 \times .0018)$
$$= .03456 + .00104 = .03560$$

LCL for $\overline{X} = \overline{\overline{X}} - (A_2 \times \overline{R}) = .03456 - (.577 \times .0018)$
$$= .03456 - .00104 = .03352$$

UCL for $R = \overline{R} \times D_4 = .0018 \times 2.114 = .00381$

LCL for $R = \overline{R} \times D_3 = .0018 \times 0.0 = 0$

Machinist Microscope

Average $(\overline{X}) = X_{total}$ divided by number of samples

Subgroup 1: $\overline{X} = .1714 \div 5 = .03428$
Subgroup 2: $\overline{X} = .1747 \div 5 = .03494$
Subgroup 3: $\overline{X} = .1723 \div 5 = .03446$
Subgroup 4: $\overline{X} = .1763 \div 5 = .03526$
Subgroup 5: $\overline{X} = .1766 \div 5 = .03532$

Grand Average $(\overline{\overline{X}}) = X_{total}$ divided by number of samples

$$\overline{\overline{X}} = .17426 \div 5$$
$$\overline{\overline{X}} = .03485$$

Range $(R) =$ Largest measurement minus smallest measurement in a subgroup

Subgroup 1: $R = .0346 - .0340 = .0006$
Subgroup 2: $R = .0345 - .0352 = .0007$
Subgroup 3: $R = .0342 - .0349 = .0007$
Subgroup 4: $R = .0350 - .0356 = .0006$
Subgroup 5: $R = .0352 - .0355 = .0003$

Average Range $(\overline{R}) = R_{total}$ divided by number of samples
$$\overline{R} = .0029 \div 5 = .00058$$

UCL for $\overline{X} = \overline{\overline{X}} + (A_2 \times \overline{R}) = .03485 + (.577 \times .00058)$
$$= .03485 + .00033 = .03518$$

$$\text{LCL for } \overline{X} = \overline{\overline{X}} - (A_2 \times \overline{R}) = .03485 - (.577 \times .00058)$$
$$= .03485 - .00033 = .03452$$

$$\text{UCL for } R = \overline{R} \times D_4 = .00058 \times 2.114$$
$$= .00123$$
$$\text{LCL for } R = \overline{R} \times D_3$$
$$= .00058 \times 0 = 0$$

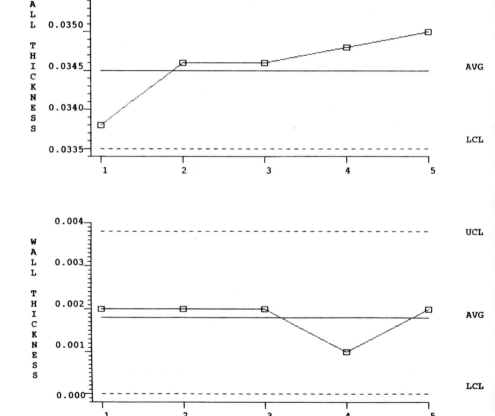

FIGURE 5.4 X bar and R chart for measurement error analysis: extruder microscope.

Step Three: Plotting the Data

Figures 5.4 and 5.5 represent the X bar and R charts for the extruder microscope and machinist microscope respectively.

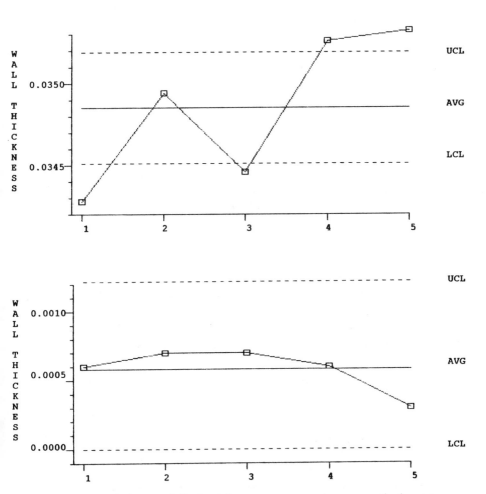

FIGURE 5.5 X bar and R chart for measurement error analysis: machinist microscope.

Step Four: Interpreting the Data—Evaluating Measurement Systems

Joe was shocked when he compared the two charts. The chart representing the extruder microscope appeared to show no signs of uncontrolled variation; the variation in the X bar chart for the machinist microscope looked totally out of control!

Mr. Wright didn't seem concerned at all. He pointed out that Joe had to stop thinking about the *production process* for a few moments because it was the *measurement process* that was being evaluated.

Comparing the Ranges Within Subgroups

In a measurement process that is **repeatable**, that is, a measurement system that combines equipment, raw material, methods, and people to yield consistent readings time after time, we would not expect to see very much difference between measurements taken on the same sample. In other words, a good and repeatable measurement process has very little range within subgroups of measurements taken on the same sample. With very little range within subgroups we can expect very narrow control limits, and, therefore, indications of what would ordinarily be considered uncontrolled variation, such as data points falling outside the upper and lower control limits.

On the other hand, a measurement device that is less repeatable has more variation. A device that has more variation will have more range within subgroups, wider control limits, and more values falling within the control limits.

In summary, two different measurement devices can be compared very quickly merely by having the same samples measured a number of times by the same people and constructing an X bar and R chart for each group of data. The same technique could be applied using one measuring device and several operators; this method, of course, would be used to determine differences in operator technique.

Joe expressed his general comprehension of the measurement error exercise, but he still had some concerns. Specifically, Joe was confused by the apparent contradiction he saw in range as a function of discrimination, and range as a function of precision.

Joe understood that an imprecise measurement system yields higher ranges within subgroups. He also understood, by virtue of the data in Table 5.2, that a measurement system with little discrimination shows greater range. Did this mean that measurement systems with less discrimination are less consistent or less repeatable?

Mr. Wright was quick to point out that Joe had to be careful not to confuse the factor of range as it applies to precision with range in discrimination. A measurement system with little discrimination will show more range in groups of measurements, but this is due to the natural rounding up or down of measurements taken. A reading of .0356 becomes .036, whereas a reading of .0352 becomes .035; the resulting range of .001 is inflated when compared to the more discriminating range of .0004. On the other hand, an imprecise measurement system will demonstrate larger ranges due to its inability to repeat. A measurement system could have enough discrimination but be imprecise; or a measurement system could have inadequate discrimination and be imprecise too.

Mr. Wright suggested that Joe think in terms of measuring the same sample over and over when considering the attribute of precision. He also suggested that Joe think of measuring samples and being able to tell the difference between the samples when considering the attribute of discrimination. In the first instance we would hope to see little or no range. In the second instance, assuming the samples are in fact different, we would want to see variation.

*Comparing the Numbers of Points Outside the
Control Limits*

A repeatable measurement system should have many, if not all, of the readings falling outside the X bar limits. Such is the case with the X bar chart using the machinist microscope. A measurement system that is not very repeatable will have many of the plotted values fall within the X bar limits. Such is the case with the X bar chart using the extruder microscope.

Notice that in both cases all the range points fall within the upper and lower control limits on the R chart. This is an indication that the measurement systems are both stable; a point falling outside the range limit would mean that the measurement system is unstable, or out of control. Such a lack of stability might result from an operator who isn't trained in the equipment or perhaps a mechanical malfunction of the measuring device.

Comparing the Standard Deviation

In addition to comparing the number of points that fall outside the X bar limits, another, more objective comparison between two stable measurement systems can be achieved by understanding the amount of standard deviation within each system.

This is easily done by dividing the already calculated R bar by the d_2 constant for the appropriate subgroup size (in this case, five).

Standard deviation for the extruder microscope is calculated to be:

$$\text{Standard deviation} = \overline{R} \div d_2$$
$$.0018 \div 2.326 = \mathbf{.0008}$$

Standard deviation for the machinist microscope is calculated to be:

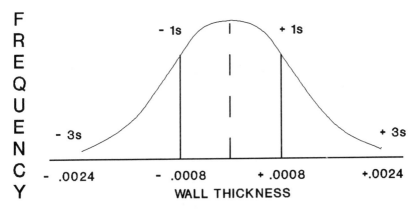

FIGURE 5.6 Standard deviation for extruder microscope ($\pm$ 1 standard deviation = .0016).

$$\text{Standard deviation} = \overline{R} \div d_2$$
$$.00058 \div 2.326 = \mathbf{.0002}$$

Figures 5.6 and 5.7 illustrate the difference in the amount of variation within the two measurement systems.

Using the curves of Figures 5.6 and 5.7, the difference between these two measurement processes can be clearly

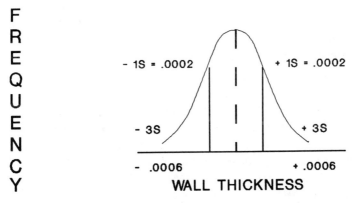

FIGURE 5.7 Standard deviation for machinist microscope ($\pm$ 1 standard deviation = .0004).

stated. We know that ± one standard deviation represents approximately 68 percent of the area under the curve. We can therefore expect that when the extruder microscope is used, the wall thickness measurement we get will be within ± .0008 of the actual wall thickness 68 percent of the time. By the same token, when we use the machinist microscope we can expect our measurement to be within ± .0002 of the actual wall thickness 68 percent of the time.

Of course, knowledge of the nature of variation and the normal curve also leads to the conclusion that, once in while, a measurement will be three standard deviations away from the actual wall thickness—or .0024 in the case of the extruder microscope and .0006 in the case of the machinist microscope.

Choosing and Implementing the Appropriate Measurement System

Joe was convinced that the machinist microscope had far more discrimination and less variation than had the extruder microscope. This was the tool he would need in order to properly measure wall thickness variation on the extrusion line. To prove it, he requested that the operator who had originally collected the data for the wall thickness control chart remeasure all of the original sixteen samples using the machinist microscope.

Figure 5.8 represents the individual control chart developed from the remeasured samples.

Joe was even more impressed when he compared Figure 5.1 to Figure 5.8. He could really see how critical the measurement system was. However, he was more than a little concerned as to management's reaction when presented with the need to equip each extruder with a $3,000 microscope.

Mr. Wright smiled when Joe expressed his concern and told Joe that he was thinking *product* control again, not *process* control. It would be expensive and impractical to place

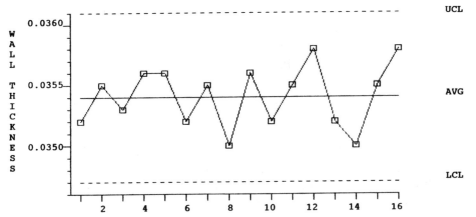

FIGURE 5.8 Individual control chart for wall thickness (measurements taken with a machinist microscope).

a sophisticated microscope, such as the one used in the measurement error analysis, on each extruder. It would not be impractical, however, to use one such microscope to measure samples from each process, and to determine the amount of wall thickness variation inherent in each process.

SETTING THE PROCESS AVERAGE

Mr. Wright referred to the extruder number six data presented in Figure 5.8 that had been remeasured on the machinist scope. He pointed out that a great deal had already been learned about the variation inherent in the extrusion process when number six was set up by the first shift operator and this particular compound extruded over this gauge and type of wire.

We already know the process average and the total amount of variation from the calculation performed to acquire the upper and lower control limits for the individual control chart.

The process average of .0354 and the wall thickness variation of ± .0007 of an inch should now be compared to the customer specification of .030 ± .004.

Mr. Wright immediately pointed out that Gigantic could expect wall thickness on this run to be as high as .036; this was not a problem as far as he was concerned, but, as Joe had already figured, TW&C was giving away a lot of material.

Consider the benefit to TW&C if the process elements were changed so as to center the process average for a wall thickness of .027 and ensure that minimum wall thickness would not go below .0263 (.027 − .0007 total variation) as long as only common cause variation is present.

Mr. Wright pointed out that he knew from his contacts with wire and cable companies that it was common practice for operators to purposely set their processes for nominal to high limit of wall thickness. Under minimum could be disastrous, but customers almost never complained about oversize wall thickness. Setting the process to yield minimum wall above nominal may have been acceptable in the days when the cost of raw materials was a much smaller portion of the overall cost of doing business, but in today's world market giving away expensive material is not good, competitive business practice.

MONITORING AND CONTROLLING PROCESS PARAMETERS

Using the control chart to understand and reduce product variation is one form of *process* control. Mr. Wright explained that the purest form of process control is attained when the tools are applied to the **process parameters**.

For instance, a *process* history can be developed over a period of time through the maintenance of control charts developed for a *product* parameter. If operators and supervisors are trained to look for, identify, and record **assigna-**

ble causes, valuable process history can be developed. This history will identify those *process* parameters, or process variables, that most often cause instability in the measured *product* parameters.

Once these process parameters are properly identified, the control chart can be reassigned from the task of monitoring and controlling the product parameter to monitoring and controlling the process parameter.

The transition from product control to process control can best be described by way of example. Consider the situation in which it has been discovered, through analysis of control chart comments, that instability of measured product is always accompanied by a corresponding rise in the extruder head pressure. This fact would then lead to a data collection exercise and the introduction of an Individual Control Chart for "head pressure."

Process average shifts detected on the head pressure control chart would be investigated and possibly found to be due to contamination buildup in the screen pack. It might be further discovered that the contamination occurs only when material from a specific supplier is introduced into the process.

The best course of action in a situation such as the one described is, of course, supplier contact directed at acquiring product without contamination. This would improve one of the elements of the process, thus improving the product.

Making a measurable improvement in a process element is process control. Once the improvement is made, the decision could be made to discontinue the data collection and control chart activity for the product parameter and/or the process parameter (head pressure).

A decision of this nature can best be made by the knowledgeable people within the organization. The decision-making process is always enhanced when people have facts to deal with, and the understanding that control charts are a means to an end. The end is world-class competitiveness, not wall-to-wall control charts that merely monitor product.

Project Team Activity

THE NEED FOR CAPABILITY STUDIES

The more Joe learned about statistical process control the more he realized how the Japanese were able to become dominant in so many manufacturing areas. Also, the more Joe learned, the more he realized how very much there was to accomplish at TW&C.

Walking through the factory the next day Joe couldn't help but notice all the measurements being taken by operators, inspectors, technicians, and engineers. There was a factory full of people making process decisions based on measuring devices that no one had studied. It was practice to calibrate all measurement devices on a periodic basis, but no one ever qualified the combination of operators and devices in the process.

Joe began to wonder how many times processes were adjusted because the operators, using the Scott Tester, determined that the elongation of the material was under minimum. It seemed that everyone more or less agreed that different operators got different readings when using the

Scott Tester, but no one had ever quantified the differences the way he and Mr. Wright had done on the microscope.

Of course, Joe could also see that processes other than extrusion would be impacted by the lack of qualified measurement systems. Lamination, braiding, irradiation, and assembly would all be affected by the use of measurement systems that did not have the proper level of discrimination or repeatability.

Joe was convinced that it was absolutely necessary to perform thorough studies on each process in order to determine whether the processes were stable and capable. He realized that TW&C had been producing product for years without knowing whether the processes were *stable*, that is, without knowing whether only normal variation was at work. But there was also no knowledge of whether any particular process was *capable* of producing 100 percent of all product within customer specifications. The only sure way of knowing if a process is capable as well as stable is to allow the process to run for a period of time without any operator adjustments. Data must then be collected and analyzed in control chart form in order to determine stability. The full range of natural process variation must then be compared to customer requirements to determine capability. Additionally, Joe now understood how important it was to fully investigate the measurement systems as a part of his studies. He concluded that **process capability studies** should be performed on all of TW&C's processes.

At the next staff meeting, Joe presented everything he had learned about statistical process control to his colleagues and the president. He went over the need to understand the process, the minimum requirement of fully understanding the measurement process, the fallacy of statistical product control, and the potential savings in material usage that could be realized if the tools were applied properly.

Joe pointed out that there was much work to be done, and he and his department could not do it all on their own.

As a matter of fact, Joe pointed out that the tools he had become acquainted with were more production and engineering tools than QA (quality assurance) tools. Joe recommended that a company-wide effort be established in order to drive the practical use of statistical process control throughout the organization. His idea was to get everyone involved to some degree, so that the people familiar with the processes could apply SPC to solve problems and to reduce process variation, scrap, and material usage.

The staff agreed that competitive pressures and customer demand didn't leave them any choice. They had to embrace statistical process control, and as long as they had to do it, they may as well get some benefit from their efforts.

It was agreed, in their usual conservative manner, to walk before running. Instead of launching into a company-wide training effort that might or might not be welcomed with open arms by everyone, the staff decided to try some "on-the-job training" in order to introduce some new people to the wonders of SPC.

IDENTIFICATION OF PILOT PROJECT AND PROJECT TEAM

It was decided to choose a particular situation that had been plaguing the organization for some time and to ask some people who were close to the problem to work as a team under Joe's tutelage. The problem chosen was excessive startup scrap on the extrusion line.

Joe left the staff meeting with the mandate to recruit a cross section of people from production and engineering and to form a team that would utilize SPC to reduce the level of startup scrap by 50 percent within a three-month period. The idea was to get people to learn by doing; people would be more inclined to learn new tools if they saw how the tools applied to their process.

The team convened and Joe began by introducing everyone to some simple concepts of variation. Joe was pleasantly surprised when it became apparent that his common-sense explanation of variation within the process elements was readily accepted by everyone on the team. Joe had been a little concerned about convincing the extruder operators that some amount of variation was perfectly normal. As a matter of fact, the operators had been experiencing variation for years, and sometimes getting blamed for excessive variation. It didn't take them long to relate to the concepts Joe was presenting.

Joe then explained the objective of the team. Joe suggested that they might consider looking into reducing start-up scrap. The team accepted this idea, and it was agreed to contact the accounting department in order to get past data on startup scrap.

PHASE ONE: THE PARETO ANALYSIS

The data received from accounting was typical of the type of data found in modern American industry—a monthly computer readout of pounds of startup scrap. (Finished scrap was measured in feet.) The information was categorized by extruder, day, job, and operator.

Joe suggested that the team work on one of the more serious problems in order that their efforts have the greatest impact. Everyone agreed with the idea of working on a serious problem, but how could anyone choose the most serious problem from the mass of computer data that was made available to the team?

Joe explained that, many times in industry, data is collected for the purposes of *accounting* for what has happened to material; it is not collected for the purpose of *solving problems*. The computer readout they had received provided valuable information for TW&C accountants; it merely had to be put into a different form for the problem solvers.

Joe suggested that they put the information into a **Pareto chart**. A Pareto chart arranges bulk information, such as is found on a computer readout, to show where the most serious areas of a problem lie.

The Pareto chart is based on what is sometimes called the "80/20 rule." Simply stated, the 80/20 rule says that in any natural situation, a majority of the results are brought about by a few of the causes. For instance, in many organizations a majority of sales come from a few of the customers; much of a department's scrap stems from only a few of the products processed through the department; the majority of absenteeism in a company is normally brought about by only a few of the people, etc. On a more personal basis, many people expend the majority of their income on a few of their creditors—car and mortgage.

In each of the examples mentioned above, the majority of the instances often represent about 80 percent of the total, whereas the few instances make up the balance—about 20 percent; hence the "80/20" rule.

Vilfredo Pareto, an Italian economist, first applied this rule to the distribution of wealth in Italy; in rough terms, he found that 80 percent of the wealth of the Italian nation was held by 20 percent of the people. Doctor Joseph M. Juran is credited with having applied this concept to areas other than economics, and with giving it the name "Pareto Principle."

Application of the **Pareto principle** is a very effective means of separating the vital few from the trivial many. In the area of problem solving, the Pareto principle can be applied to identify the area of the problem that should receive most of the attention.

Joe and the team spent some time and converted the data on the computer readouts to Pareto chart form. They created two charts: one representing startup scrap by extruder and one representing startup scrap by shift. Figures 6.1 and 6.2 illustrate the two charts.

The team could see that it would make much more sense

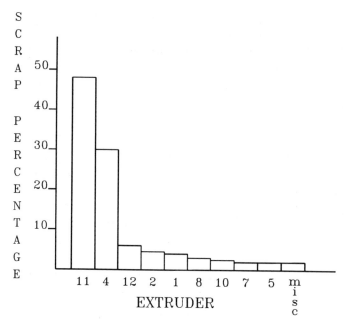

FIGURE 6.1 Pareto chart for startup scrap by extruder.

to apply their efforts to extruder #11 or #4 because these two extruders represented the vast majority of startup scrap. The team could also see that there wasn't much difference between the amount of startup scrap between the three shifts. This latter revelation came as a surprise to a number of team members; many people just "knew" that the third shift produced more startup scrap than the other two shifts. Joe quietly pointed out that one of the advantages of examining statistical data is that we deal with facts, not feelings. Obviously, the feeling that the third shift produced more startup scrap was not accurate. The fact is that each shift produced about the same amount. This fact is very important because the identification of a nonproblem is as important as the identification of a real problem.

Everyone agreed to single out extruder #11 and attempt

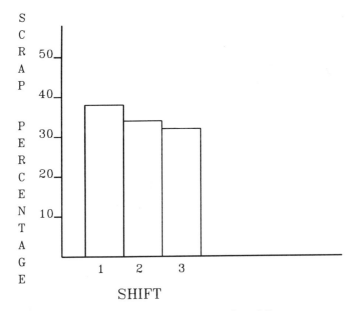

FIGURE 6.2 Pareto chart for startup scrap by shift.

to identify the causes for what appeared to be excessive startup scrap.

The team met once with each of the operators assigned to extruder #11. The operators were unanimous in identifying the cause of the problem: *Extruder #11 was used for short runs, and the family of compounds that were run on #11 required a minimum internal specification of 200% elongation.* The minimum elongation specification was so difficult to attain that sometimes the startup scrap equalled or exceeded the good product run.

Several members of the team were frustrated; how could the problem be solved without extending the length of the runs or changing the internal elongation specification? Neither option was within the power of the team.

Joe indicated that a problem first has to be identified before it can be solved. The problem is not the specification or the length of the run; the problem is excessive startup scrap. Changing either the length of the runs or the specification would be like making the patient comfortable instead of curing the illness.

Joe pointed out that the team was in the problem-solving business, not the patient comfort business. They had to identify the real problem and work on it.

PHASE TWO: THE CAUSE-AND-EFFECT ANALYSIS

After Joe led the team through a **cause-and-effect exercise**, all agreed that the real problem was the length of time it took operators to attain acceptable elongation.

At their first team meeting, Joe introduced the nature of variation, the definition of stability, and the definition of capability. The team discussed the possibility that operators were having difficulty attaining the elongation spec because the product parameter "elongation" had too much variation. Someone suggested that the problem of attaining the elongation spec in a timely fashion might stem from differences between one startup batch and another. In statistical process control terms, the team was discussing the capability and the stability of the process.

Figures 6.3 and 6.4 graphically represent the two concepts that were discussed.

Lack of capability in a situation such as minimum elongation requirements might result from a natural distribution, in which the average could not be driven beyond a certain limit. Figure 6.3 illustrates a natural distribution ranging from 175% to 225% with a process average of 200%. This process would be considered incapable of achieving 200% minimum elongation all the time if, because of the material's physical limitations, the average could not be shifted higher than 200%.

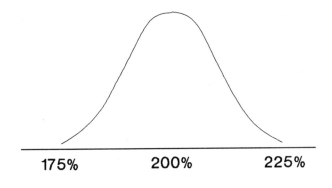

ELONGATION

FIGURE 6.3 Excessive variation compared to specification (200%). Process is not capable.

PHASE THREE: PROCESS CAPABILITY STUDY

Joe explained to the team that the only way to identify whether a process is stable and/or capable is to perform a properly run process capability study. Such a study would

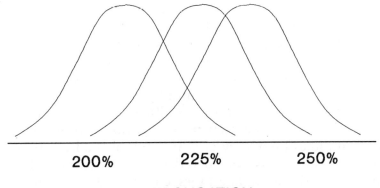

ELONGATION

FIGURE 6.4 Excessive variation due to raw material batch variation. Process is not stable.

provide the team with insight on the effects of normal variation due to equipment, raw materials, methods, and people.

He further explained that for a process capability study, it is necessary to allow the process to run for a period of time without making any adjustments. If all operator-induced instability is removed from a study, the data will reflect only that variation resulting from the equipment and the raw material. Joe assured everyone that the most difficult part of a process capability study is convincing the operators to leave things alone once they have things running the way they want them.

Measurement Error Analysis

Before any process capability study could be performed, however, it was absolutely necessary to perform a measurement error analysis. Joe knew that one of the primary reasons for running a process capability study is to understand how much normal variation exists in the process. In order to understand how much normal variation exists in just the process, he would have to know how much of the recorded variation was due to measurement error. The Scott Tester used by the operators had to be studied.

Because elongation is a destruct test, some special provision had to be made with regard to the sample preparation. One sample was taken from five individual reels of the product to be studied; each sample was five feet long. Each five-foot sample was cut up into fifteen equal lengths.

On the assumption that the elongation within each five-foot sample did not vary greatly, each of the five groups of fifteen samples was considered to be one sample. In this way, three different operators could measure "the same sample" on the Scott Tester five times each.

Table 6.1 represents the readings from each of the three operators.

TABLE 6.1 Sample Elongation Measurements Taken by Three
Operators Using the Scott Tester

| | OPERATOR A | | | | |
	Reel 1	Reel 2	Reel 3	Reel 4	Reel 5
	200%	250%	180%	250%	200%
	225%	275%	180%	250%	225%
	250%	225%	220%	225%	200%
	250%	250%	210%	275%	200%
	225%	225%	210%	250%	250%
$\overline{X}$	230%	245%	200%	250%	215%
R	50%	50%	40%	50%	50%

| | OPERATOR B | | | | |
	Reel 1	Reel 2	Reel 3	Reel 4	Reel 5
	200%	250%	200%	250%	200%
	225%	250%	225%	275%	250%
	225%	250%	225%	275%	225%
	225%	225%	200%	250%	225%
	200%	225%	210%	250%	250%
$\overline{X}$	215%	240%	212%	260%	230%
R	25%	25%	25%	25%	50%

| | OPERATOR C | | | | |
	Reel 1	Reel 2	Reel 3	Reel 4	Reel 5
	225%	250%	200%	250%	200%
	225%	225%	225%	225%	225%
	250%	275%	180%	250%	190%
	250%	275%	250%	200%	225%
	210%	225%	250%	250%	225%
$\overline{X}$	232%	250%	221%	235%	213%
R	40%	50%	70%	50%	35%

Joe explained to his team that the traditional X bar and
R charts would be used to evaluate the measurement process. Figures 6.5a and 6.5b represent the X bar and R control charts developed from the measurements taken by the

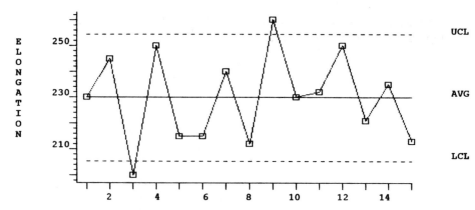

FIGURE 6.5a X bar chart for elongation (measurements taken by three operators using a Scott Tester).

three operators using one Scott Tester. (Please refer to the Appendix for formulas and factors.)

There was absolutely no way this measurement system, in its present state, could be used to determine the capability of the process. The standard deviation, which was cal-

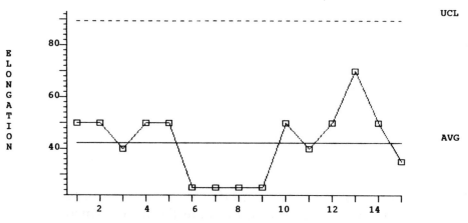

FIGURE 6.5b R chart for elongation (measurements taken by three operators using a Scott Tester).

culated as follows, indicated that there was far too much error.

The average range for all of the readings was 42%. One standard deviation was therefore equal to 18.1%, as calculated below:

$$\overline{R} \div d_2 = 42\% \div 2.326 = 18.1\%$$

This meant that 68 percent of all the readings taken by these three operators would be within ±18.1% of the actual value of elongation.

In addition, Joe explained that Mr. Wright had said that the $\overline{X}$ chart was a good indicator of excessive measurement error. When a majority of values on the $\overline{X}$ chart fall inside the control limits, it is a very good indication that the measured values are being dominated by error.

The team was surprised at the amount of error, especially because operators and supervisors were in the habit of making process decisions based on the readings from the Scott Tester. Joe pointed out that this amount of error in the measurement system could be a significant part of the problem they were working on.

It certainly seemed logical that an operator using this measurement system could be fooled into thinking that a process was producing product well below the elongation specification when, in fact, the product was within the limit, or vice versa. Not only would this situation result in excessive adjustments in order to bring the elongation into "specification," but it could also result in the entire run being below the minimum.

Joe could remember a number of times in his career when the operators' records showed product parameters to be well within specification during the production run, and yet the product would fail at inspection. How many times, he wondered, had operators been unfairly blamed for not doing their job when, in fact, everyone was fooled by the measurement system.

Interviewing the operators who had participated in the measurement error analysis helped the team understand why there was so much error. The ruler used to measure the benchmark was a standard grade-school ruler. The ruler was too long to fit between the cross heads, so the operators were forced to take a reading while the ruler was on an angle. This was made worse by the fact that the operators were all different heights. Also, there was very little standard procedure with regard to the preparation of the sample. Some operators marked the sample with a narrow ball-point pen, whereas others used a broad grease pencil. Everyone used the same beat-up wooden ruler to benchmark the inch. Further, the lighting was poor in the area of the Scott Tester, and none of the operators had ever received training on the use of the equipment from anyone other than another operator.

Joe realized that the process capability study could not continue until the measurement process had less variation. He therefore instituted training sessions for the operators and provided them with standardized written procedures to follow when taking measurements. Additionally, he provided them with better tools, including a short customized ruler, and improved the poor lighting condition that had adversely affected the Scott Tester readings. Shortly thereafter, the measurement error analysis was performed again.

Table 6.2 represents elongation readings taken by the same operators after the training sessions.

Figures 6.6a and 6.6b represent the revised X Bar and R charts.

Clearly the measurement error had been considerably improved; one standard deviation was now equal to only 7.3%:

$$\overline{R} \div d_2 = 17\% \div 2.326 = 7.3\%$$

Now the team could proceed with the process capability study confident that the actual elongation data would not

TABLE 6.2 Repeated Elongation Measurements Taken Under Improved Conditions

	OPERATOR A				
	Reel 1	*Reel 2*	*Reel 3*	*Reel 4*	*Reel 5*
	200%	250%	200%	250%	200%
	215%	260%	200%	270%	200%
	200%	250%	200%	250%	210%
	200%	260%	220%	250%	200%
	215%	260%	200%	270%	210%
$\overline{X}$	206%	256%	204%	258%	204%
R	15%	10%	20%	20%	10%

	OPERATOR B				
	Reel 1	*Reel 2*	*Reel 3*	*Reel 4*	*Reel 5*
	215%	250%	200%	250%	200%
	220%	250%	215%	250%	200%
	215%	270%	200%	240%	200%
	220%	250%	200%	250%	220%
	220%	250%	215%	260%	200%
$\overline{X}$	218%	254%	206%	250%	204%
R	5%	20%	15%	20%	20%

	OPERATOR C				
	Reel 1	*Reel 2*	*Reel 3*	*Reel 4*	*Reel 5*
	200%	250%	210%	250%	200%
	190%	250%	200%	270%	210%
	200%	270%	215%	250%	215%
	200%	250%	200%	250%	220%
	210%	270%	200%	250%	200%
$\overline{X}$	200%	258%	205%	254%	209%
R	20%	20%	15%	20%	20%

be as greatly affected by the measurement system. Granted, there was still some degree of error present, but the error was less and they knew what the error was.

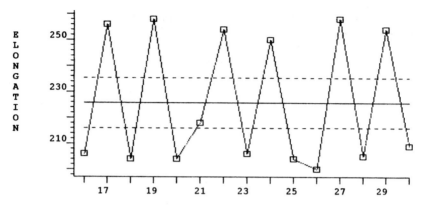

FIGURE 6.6a X bar chart showing remeasured elongation data.

Individual Control Chart

Armed with the new and improved measurement system, the team set about to run the process capability study.

The team set up the equipment according to standard practice, made certain enough raw materials were on hand to support the study, waited until the operator was satisfied with the quality of the product being produced, and then

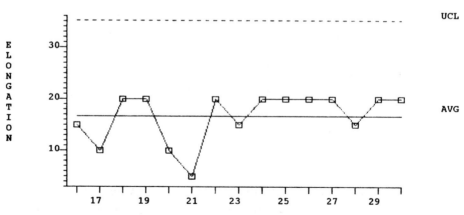

FIGURE 6.6b R chart showing remeasured elongation data.

TABLE 6.3 Data Collected for
Capability Study

ELONGATION MEASUREMENTS	
225%	215%
210%	230%
215%	215%
230%	205%
215%	220%
235%	230%
225%	210%
205%	220%
215%	225%
230%	210%
225%	230%
215%	210%
200%	

began to collect data. The team ensured that no operator adjustments were made during the entire capability study.

Table 6.3 illustrates the data collected during the run. Figure 6.7 illustrates the individual control chart derived from the data.

Joe reminded everyone that the readings taken for the capability study would be affected by the measurement error. A reading of 210% might be higher or lower than the

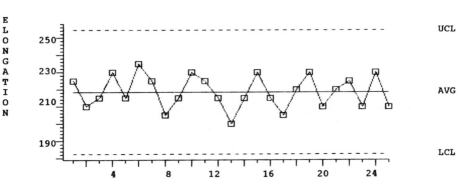

FIGURE 6.7 Individual control chart for elongation.

actual elongation of the sample. Knowing that one standard deviation of the measurement error was equal to 7.3%, it would be necessary to exercise extreme care when samples were found to be close to the minimum requirement. Perhaps it would be prudent to measure low samples on the more precise Instron. Of course, a measurement error analysis would first have to be performed on the Instron.

The main point impressed upon the entire team was the fact that careful thought had to precede any process adjustment.

Results of the Capability Study

The process, without operator adjustments, is stable and is capable of producing product within ±36% centered about some process average elongation.

The team members were feeling pretty good about their accomplishments so far. They had qualified a measurement system, and had developed a Scott Tester training program complete with written procedures. Also, they had determined that extruder #11 could produce at least one family of product with consistent elongation. Finally, they had demonstrated to the operators that there was no need to constantly make adjustments to the process in order to maintain elongation.

However, the team had not accomplished any measurable gain with regard to their stated goal of reducing start-up scrap.

PHASE FOUR: PROCESS AVERAGE SETTING PROCEDURE

At this point a team member asked a very interesting question:

- Now that we understand how much the elongation does vary, and we have a good way to measure it, how does the operator know where the process average is at setup?

- Let's say that the process average, at initial setup, happens to be centered at 190%, and it's just my luck that my first reading is taken when the process is yielding product at 210%. My reading is going to tell me that I'm making good product, but as we now know, some of my product will be under low. I'll be making scrap and not know it.

- I'm not always going to be lucky enough to start out with the process average exactly where I want it, and while I'm adjusting to get what I want, I'm making scrap. Control charts tell me if I'm stable over a period of time, but they don't help me understand where I am at the beginning of a run. It looks to me like we are right back where we started from—making scrap while we hunt for the right settings.

Joe had to admit that he had not considered this limitation of the control chart tool. Joe promised to bounce this problem off of Mr. Wright to see if he had any ideas.

Mr. Wright came through as usual. Not only did he have the answer to the problem, he faxed it to Joe that very afternoon. Joe pulled together some overheads of the information he had received from Mr. Wright and made a presentation to the team the next day.

Joe explained that the answer to the question lay in the basic understanding of variation and stability. Any sample taken from a stable process, because of the laws of probability, has about a 68-percent chance of coming from the area between plus or minus one standard deviation. This is because about 68 percent of everything under the curve falls between plus or minus one standard deviation.

Figure 6.8 illustrates the approximate proportion of

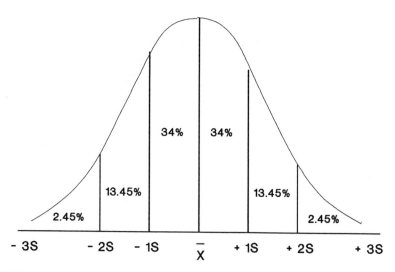

FIGURE 6.8 Approximate proportions under the normal curve.

product as it falls within the normal curve. (Notice that the sum of all percentages does not equal 100 percent; this is due to a statistical theory that will not be covered in this text.)

Joe didn't go into the details of the statistical theory that is the basis for the following setup criteria. He suggested that the operators simply accept the fact that some very clever statisticians had done a lot of probability calculations to give them a tool that would help them set the process average where they wanted it.

Joe used an example that defined product specification at 245% ± 45%. The process standard deviation for the control chart information in Table 6.3, calculated with the formula:

$$\text{Standard deviation} = \overline{R} \div d_2$$

was 12%. He then gave the operators the criteria for the readings at setup, as follows.

If, at initial setup, an operator's first reading falls between nominal and ± 1.44 standard deviations (245% ± 17.3% for this example), the operator is 68 percent confident that the process average is within 1.44 standard deviations of nominal. When this criterion is met, the decision should be *do not adjust*.

If the average of the operator's first and second readings falls between nominal and ± 1.02 standard deviations (245% ± 12.2% for this example), the operator is 68 percent confident that the process average is within 1.02 standard deviations of nominal. Again, when this criterion is met, the decision should be *do not adjust*.

The criteria or prescribed limits for successive readings are given in Table 6.4.

As long as each of the averages falls within the nominal plus or minus the required distance in standard deviation units as specified in the table, the decision should be not to adjust. If the average of one group falls outside the prescribed limits, then the operator should make an adjustment and start over.

The team agreed that this technique would assist in helping the operators make decisions with regard to initial setup.

TABLE 6.4 Criteria for Operator Readings at Setup

Reading	*Prescribed limit*
1	Nominal ± 1.44 std. dev.
avg. of 1 and 2	Nominal ± 1.02 std. dev.
avg. of 1–3	Nominal ± .84 std. dev.
avg. of 1–4	Nominal ± .72 std. dev.
avg. of 1–5	Nominal ± .65 std. dev.
avg. of 1–6	Nominal ± .59 std. dev.
avg. of 1–7	Nominal ± .55 std. dev.
avg. of 1–8	Nominal ± .51 std. dev.

PHASE FIVE: PILOT PROJECT CONCLUSION AND RECOMMENDATIONS

The team reviewed everything they had learned about the process, and they were amazed at how much TW&C had been operating in the dark. Before the team activity, no one knew for sure which jobs produced the majority of startup scrap; everyone had their own idea, however. Also, no one had ever considered the idea that variation in measurements of elongation could lead to excessive adjustments, scrap, and processing problems. Last, but far from least, before the team activity nobody had any idea how much variation existed in the product parameter "elongation," and the team now understood how critical this knowledge was to achieving quick setup.

The team pulled together all the information on the process and made recommendations to the staff as to how to go about reducing the amount of startup scrap on the extrusion line. The recommendations included introducing control charts for the important parameters and using the process average setting procedure.

The team further recommended that other extruders be studied in the same way that #11 had been studied, but by other people. Everyone on the team agreed that they had learned more process facts in a few short weeks than they had in years of operating an extruder; other people should have the same opportunity to learn the facts so that they could begin to base their actions on facts rather than sentiments.

Finally, the team recommended that some means be introduced to measure the progress of startup scrap reduction—some means other than the computer readout that was developed for accounting once a month. The team suggested that a Pareto chart be maintained on the extrusion line; the purpose of the chart would be to record the *estimated* dollar amount of startup scrap on a weekly basis. The team felt that if everyone could see how much the company was losing due to startup scrap, operators would be more inclined to take precautions and use the tools that were being developed to help reduce the scrap.

The team had learned some important lessons with regard to measurement; it was time to pass these lessons on to others, because it was in everybody's best interest to reduce scrap and make the company more competitive.

The more competitive the company, the more secure would be everyone's job.

Use of the P Chart

CONTINUOUS IMPROVEMENT

The statistical process control effort was well under way at TW&C. More importantly, SPC had become part of an over-all continuous improvement effort that was systematically addressing each process in order to reduce variation. Operators, supervisors, technicians, engineers, and office personnel were all being drawn into team activity. Everyone was beginning to appreciate the power of data collection, because they were seeing positive results. Furthermore, some of the old battles were no longer being fought because people were no longer relying on sentiment to draw lines in the dirt.

Management was beginning to form the habit of requiring data before making decisions. For instance, data was required before making decisions regarding the purchase of replacement capital equipment. Management was now requiring that a capability study be performed on the old equipment in order to determine if, in fact, it needed replacement.

Some department heads were now voluntarily approaching Joe to ask for his help in understanding the variation within their processes. Although they may not have used that exact terminology, their needs were clear.

With all the positive activity going on, Joe couldn't help but notice that one area of the plant was still devoid of any continuous improvement effort. The cable assembly department seemed to be totally out of the mainstream of activity.

In conversations with the cable assembly department head, Joe was told that all of the cable assemblies produced at TW&C were required by customer contract to be sorted 100 percent. There was no sense in introducing SPC to an area where all product underwent 100 percent inspection. Joe was not entirely sure that he agreed with the cable assembly department head; however, he wasn't sure enough of his grounds to take issue. Joe beat a hasty retreat in order to think this one out.

In the quiet of his office Joe began to do a bit of research on what SPC tool might be useful in an environment such as the cable assembly department. The manufacture of cable assemblies was not at all like extruding or bunching wire. As a matter of fact, the type of defect produced wasn't even the same. The majority of defects in the department, Joe knew from his early days as a production supervisor, were cosmetic. The majority of defects in most of the other operations were something you could measure: low wall, undersize diameter, lack of concentricity, etc.

Joe needed a way to statistically understand the variation of cosmetic defects. An article in a trade journal reminded Joe of a technique he had learned earlier in the year at the SPC seminar he had attended. The P chart and the C chart were two tools he had almost forgotten, but both were designed to help people control the variation of nonmeasurable defects such as cosmetic defects—better known as **attribute** defects.

VARIABLE DATA VERSUS ATTRIBUTE DATA

The X bar and R charts and individual charts, the control charts discussed thus far, deal with variable data. **Variable data** is data that can be measured and described with numbers.

Attribute data, that type of data best addressed by means of a P chart or C chart, describes whether a condition is present or not present. Following are some examples of attribute data.

- A product is either black or it is not black.
- A product either fits into a gauge or it does not fit into a gauge (go/no go).
- A product is either visually acceptable or it is not visually acceptable.
- A product is either defective or it is not defective.

None of the above examples lend themselves to being measured by means of a device that offers a numerical description of its condition.

Measurement Tools for Attribute Data

Both the P chart and the C chart contribute to the understanding of attribute data: both deal with values that are not numerical. As with the other tools of SPC, the P chart and the C chart can be applied only in certain circumstances.

The C Chart

A brief review of his notes concerning the **C chart** reminded Joe that C chart technology would be applicable in a process in which the complexity of the product might result in a number of defects on the finished unit. For example, in the production of automobiles, all the defects combined would not cause the manufacturer to scrap any one car. However, over the life of the model, it would benefit the manufacturer to understand what constitutes a normal number of defects per automobile and to take steps to reduce the average number of defects per unit as well as the amount of variation between the anticipated maximum and minimum number of defects per unit.

Thus, the C chart would only be of benefit to the cable assembly department if the supervisor is interested in understanding the variation of the number of defects on each component. As a practical matter, a cable assembly is bad whether it has one defect on it or ten. Therefore, it is important to the success of the department to know how much variation there has been in the number of defective components being produced, as opposed to the number of defects produced on each component. Joe realized that the C chart would not be applicable in the cable assembly department.

The P Chart

The **P chart,** on the other hand, is a tool designed to provide information on the quantity of defective components in a production run. This was the SPC tool that Joe could use to statistically understand the variation of cosmetic defects in the cable assembly department. When working with a P chart, you are generally calculating what has been termed the "percent defective." The data in most P charts is expressed in terms of percentages. An example that utilizes the P chart is discussed later in this chapter.

Variation Within Attribute Data

People in American industry have been fooled by variation within attribute data in much the same way that they have been fooled by variation within variable data. This is in large part due to the fact that attributes vary in much the same way as do dimensions.

Consider a situation in which seven half-dollars are lined up on a table. When lying flat, each coin can exhibit either a head or a tail; a coin could not, for instance, be showing .66 of a head—a perfect example of attribute data. In flipping the coins, the probability of all seven coming up heads is very remote. As a matter of fact, the **probability** of such an occurrence is 1/128; or, in 128 tries, all seven coins would come up heads only once. The probability of getting six heads, five heads, four heads, three heads, two heads, one head, and no heads is graphically illustrated in Figure 7.1.

Figure 7.1 looks suspiciously like a normal distribution. The natural laws that dictate the variation of attribute data are not very different from the natural laws that dictate the variation of variable data.

If money were to be bet on a random toss of seven coins, smart money would be put on the chance that three or four

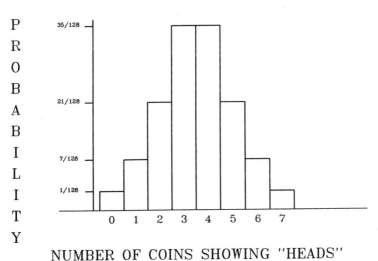

NUMBER OF COINS SHOWING "HEADS"

FIGURE 7.1 Probability in a coin toss (7 coins tossed).

heads would appear. This is because smart money knows that the probability of getting three or four heads is far better than for any other combination. In other words, *on the average* we could expect three or four heads to appear far more often than any other combination; the other combinations will appear, but not as often.

In the same way, in a number of samples randomly chosen from a manufacturing process that is only being affected by common cause variation, the probability of finding the *average* percent defective is greatest. Of course, the probability of finding a much smaller or much larger percent defective also exists.

P CHARTS IN MANUFACTURING

Taking a sample from a manufacturing process can be very misleading, for all of the reasons already mentioned. Operators or roving inspectors can sample a process and, because of the laws of probability (luck of the draw), walk away with the perception that the process is yielding a very small percentage of visual defects. The material may receive a very different review in final inspection, however, again because of the laws of probability. Who is right? The operators and roving patrol, or the final inspectors?

Much of the friction between some production and quality departments comes from the misunderstanding stemming from false perceptions based on sample plans.

P charts offer a means to determine the limits of expected percent defective when only common cause variation is present. This is an extremely important tool because, as in the case of individual charts, high and low readings that are not outside calculated limits will not be cause for alarm. In devising a P Chart for attribute data, you will go through similar steps to those used for the X bar and R charts and the individual charts.

CONSTRUCTING A P CHART

Step One: Collecting Samples and Measurements

Consider a cable assembly process that yields, on the average, 8 percent visually defective product due to scratches. This is not too farfetched; many American companies build into their product cost a 10-percent scrap rate. Customer requirements mandate a 100-percent sort of all lots. A series of thirty lots were processed in this manner.

For each lot, the number of defects due to scratches was recorded and the percentage of defects in the lot was then calculated. Working in percentages allows an accurate comparison to be made when lot sizes vary. To arrive at the **percent defective** figure, the following formula is used:

percent defective (P) = (number of defective pieces divided by sample or lot size) times 100%

For example, in the first lot, 150 components were checked, and 11 of these contained scratch defects. The percent defective calculation for this lot is:

$$11/150 \times 100\% = 7.3\%$$

In the second lot, 170 components were checked, and 5 contained scratch defects. The calculation for this lot is:

$$5/170 \times 100\% = 2.9\%$$

Table 7.1 represents the results of the inspection for all thirty lots. It is important to note that each lot is a sample chosen from a continuous production line that has a constant level of defects. When we inspect a lot we are in fact taking a random sample from the entire process output, and the percent defective in any one lot is a result of that lot being a random sample from the entire **population**.

TABLE 7.1 Defective Cable
Assembly Data From 100 Percent
Inspection

Lot size (n)	# def	% def (p)
150	11	7.3
170	5	2.9
145	17	11.7
180	5	2.8
134	10	7.5
157	15	9.6
186	9	4.8
143	14	9.8
140	6	4.3
175	15	8.6
143	17	11.9
184	18	9.8
144	15	10.4
142	10	7.0
176	9	5.1
146	11	7.5
155	18	11.6
178	11	6.2
180	21	11.7
154	8	5.2
154	12	7.8
178	26	14.6
157	12	7.6
178	21	11.8
143	2	1.4
176	22	12.5
148	10	6.8
178	21	11.8
175	11	6.3
178	14	7.9

Step Two: Calculating the Average Percent Defective

The average percent defective (P bar or $\bar{P}$) for all thirty lots is calculated by dividing the total number of rejects by the total number inspected, then multiplying that number by 100% so that the figure is expressed in a percentage.

$$\overline{P} = (\text{total rejects} \div \text{total inspected}) \times 100\%$$
$$\overline{P} = (396 \div 4847) \times 100\%$$
$$\mathbf{\overline{P} = 8.17\%}$$

Step Three: Calculating the Upper and Lower Control Limits

As with the other control charts, there is a standard formula for calculating the upper and lower control limits for the P Chart. The two formulas follow. Note that "n" equals the **sample size**.

$$UCL = \overline{P} + 3\sqrt{\frac{\overline{P} \times (100 - \overline{P})}{n}}$$
$$LCL = \overline{P} - 3\sqrt{\frac{\overline{P} \times (100 - \overline{P})}{n}}$$

As long as the maximum and minimum lot sizes are within 25 percent of the average lot size, the average lot size can be used for the sample size (n). The average lot size is calculated by dividing the total number of samples by the total number of lots. In this example the average lot size is 161.6:

$$n = 4847 \div 30 = 161.6$$

As you will note from the control limit formulas, the arithmetic for the upper and lower control limits can be shortened merely by adding and subtracting from **P** bar the nonvariable term:

$$3\sqrt{\frac{\overline{P} \times (100 - \overline{P})}{n}}$$

Now that you have both the average percent defective ($\overline{P}$) and sample size (n), the arithmetic for the nonvariable term can be performed as follows:

$$3\sqrt{\frac{8.17 \times (100 - 8.17)}{161.6}} = 3\sqrt{\frac{8.17 \times 91.83}{161.6}}$$

$$= 3\sqrt{\frac{8.17 \times 91.83}{161.6}}$$

$$= 3\sqrt{\frac{750.25}{161.6}}$$

$$= 3\sqrt{4.64}$$

$$= 3 \times 2.15$$

$$= \mathbf{6.45\%}$$

The upper and lower control limits can now be easily calculated as follows:

$$\text{UCL} = \overline{P} + 6.45\%$$
$$= 8.17\% + 6.45\%$$
$$= \mathbf{14.62\%}$$

$$\text{LCL} = \overline{P} - 6.45\%$$
$$= 8.17\% - 6.45\%$$
$$= \mathbf{1.72\%}$$

Step Four: Plotting and Interpreting the Data

Figure 7.2 represents the P chart for the data found in Table 7.1. If we make the assumption that only common cause variation was present during the time the data was collected, the upper and lower control limits represent the expected limits of percent defective during normal operation.

In other words, as long as the lots are running between 1.7 percent and 14.6 percent defective, everything is normal.

Management might have some difficulty accepting this condition, but then it would be up to management to take steps to reduce P bar as well as the amount of variation. One means by which P bar and the width of the control

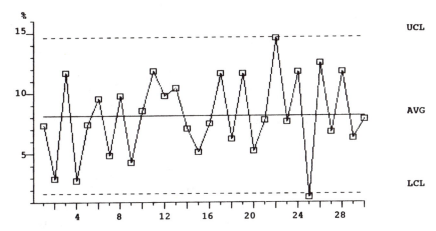

FIGURE 7.2 P chart for cable assemblies.

limits can be improved is through the active and ongoing use of the P chart.

APPLICATION OF THE P CHART

TW&C would benefit greatly if the cable assembly department operators would record each lot percent defective on a P chart, instead of merely recording the number defective on some obscure inspection sheet. If the P chart was maintained, management would have a history of the process instead of raw numbers on a computer readout at the end of the month.

Consider the advantage that TW&C would have if a P chart indicates a number of consecutive lots all yielding a very low percent defective—well below P bar, approaching zero. This might be an indication that something out of the ordinary was going on in the process. In this context the "something out of the ordinary" is producing a desirable result: less scrap. Operators and supervisors could stop and try to identify what conditions prevailed to yield such a welcome result.

Joe decided that he had come across another instance in which operators and supervisors were doing a great deal of work but not getting all the benefit. As long as the customer was requiring 100-percent sort, TW&C might just as well learn something about the process variables while they were at it.

For the cable assembly department, Joe was convinced that the best way to learn about the process was to implement P charts on a few of the most serious attribute defect/product combinations. Once again, Joe decided that the best way to convince the supervisor to implement a new technique was to prove that there would be a direct benefit to the department.

Joe knew that personnel turnover was a problem in the cable assembly department. The supervisor was always complaining about difficulty with new people learning the visual sort standards. Every time a new person started, there was a period during which an excessive amount of good product was scrapped and/or an excessive amount of bad product was passed. Also, there was a problem of visual standards slipping among the older operators.

Joe sat down with the supervisor and presented the P chart as a training tool. He suggested that the percent defective plotted on the charts by the operators would be a very good indication of their adherence to standards. A series of very low values might indicate that a new employee was not being critical enough and passing bad product; a series of high values might indicate that an operator was being too critical and rejecting too much.

Joe pointed out that process changes made by engineering could also be evaluated by means of the P chart. This really grabbed the attention of the supervisor, who was always complaining that the engineers were constantly making "process improvements" that only made the scrap situation worse.

Joe pointed out that any changes to the process—good or bad—could be factually evaluated by means of the P chart. If after a process change P bar was decreased and the control limits became narrower, then the organization could point to an improvement. On the other hand, if the control limits got wider and/or P bar increased, the engineers would have proof that the change was not for the better.

The cable assembly department supervisor agreed to implement, on a trial basis, a series of P charts. Joe knew that once the P charts were introduced, they would prove their own worth and become part of department procedure.

SUMMARY

As Joe walked around the factory, he couldn't help reflecting on some conversations he had had several months ago with Mr. Wright of Gigantic Wire Using Corporation. Mr. Wright had, in essence, pointed out that you can lead a horse to water but you can't make him drink—an old adage but one that totally applied to introducing an organization to statistical process control. All the number-crunching seminars in the world can't compete with showing the local work force how SPC would benefit them in the way they do their job.

Convincing people that they should even be involved in improvement can be a chore. In many companies, operators gave up offering ideas to improve the process a long time ago; no one ever listened anyway. The best way to get people involved is to challenge them to improve their processes and then give them the tools to factually prove or disprove their ideas. Joe knew that initial skepticism would soon give way to enthusiasm when the operators realized that everyone, including management, listens when opinions are turned into fact by means of data collection and SPC.

Of course, all the local work force enthusiasm and SPC knowledge is totally ineffective if management isn't committed to organizing for improvement. Statistical process control knowledge without a formal structure to apply it is like being all dressed up and having no place to go. Joe knew from experience that a team of people given an objective by management and trained in SPC and problem solving is an awesome tool. Joe liked to tell people that a project was a problem scheduled for solution.

Best of all, for the first time in his life Joe felt like a quality professional. He enjoyed his new role of helping provide production and engineering with the tools to do the job right the first time.

Because everyone was embracing SPC and the continuous improvement philosophy, Joe wasn't spending all of his time running around engaged in crisis management. Nor was he spending his time attempting to explain to management why the scrap figures were so high; he never could understand why they always asked him anyway. Joe was spending more and more of his time refining his understanding of SPC and continuous improvement. There was much more to learn, but he was no longer apprehensive about trying new statistical ideas and techniques because he had found a number of different techniques that worked very well. Also, he had come across an encouraging quote by Walter Shewhart, the father of industrial statistical process control: "The fact that the criterion which we happen to use has a fine ancestry of highbrow statistical theorems does not justify its use. Such justification must come from empirical evidence that it works."

Meeting the Challenge

The story of Joe and TW&C is a composite of a number of actual experiences encountered by the author. The personalities, attitudes, and situations described are, in the experience of the author, commonplace in American industry.

America has begun to lose its ability to compete in the world marketplace. We risk becoming a second-rate industrial power, not because of a lazy or disinterested work force, but primarily because American management has relied upon old and outdated methods.

In July of 1989, the Massachusetts Institute of Technology (MIT) Commission on Industrial Productivity reported that the nation was in far more serious trouble with regard to industrial performance than the national leadership was admitting.

Among the recommendations of the Commission was the following: American manufacturers must do a better job of training the work force as well as expanding the scope of the average worker's responsibilities. Implied in this recommendation is the observation that, in the past, American companies have not done enough in the area of training and educating the American worker.

The story of Joe and TW&C attempts to illustrate how far a little bit of education can go to make an organization more competitive. In our story, Joe was able to accomplish a great deal in a relatively short period of time. In real-world SPC implementation, however, the same degree of accomplishment generally takes much longer. Ironically, the delays in implementation stem more from management attitudes and lack of understanding than from resistance on the part of the work force.

An effective SPC effort is often delayed, and sometimes never brought about, in cases where management is interested in SPC solely because implementation has been mandated by a major customer. Customer mandates to embrace SPC are sometimes taken too literally. Management sometimes overlooks the proposition that SPC is only one of several tools required to sustain long-term improvement and competitiveness; and it is long-term improvement and competitiveness that is at the heart of the customer's request. As one of several tools, SPC must be used in conjunction with other tools to be effective. When the tool of SPC is introduced by itself, and is subsequently deemed to be less than totally effective, disappointment and delays set in.

A quality plan such as Ford Motor Company's Q-1 is a good example of how this phenomenon comes about. In order to do business with any division of Ford Motor Company, an organization must be qualified as a Ford Q-1 Supplier. This qualification involves a very thorough survey by a Ford Supplier Quality Assurance engineer. The intent of the survey is to determine the ability of a supplier company to continuously improve its processes in order to supply Ford with the highest quality product at the lowest cost. Although a company's entire attitude toward quality is surveyed, an emphasis is placed on the practical application of statistical process control, because SPC is one of the primary tools used for process improvement.

It is probably this emphasis on SPC, by Ford and other major companies, that causes supplier companies to believe that SPC implementation is the goal rather than a means of achieving the actual goal of continuous improvement. When SPC implementation is viewed as the goal rather than as the means to the goal, management falls into the trap of thinking implementation is a very focused affair, one that does not affect the entire company.

This limited view of what SPC is all about sometimes leads to internal instructions to the quality control department to establish an SPC program. Instructions such as these indicate a lack of understanding on the part of management, because they imply that SPC is just a tool of the quality control department. Also, such instructions indicate a total lack of understanding regarding the amount of training required throughout the organization—not only in basic SPC tools and techniques, but also in project team activity and non-numerical problem-solving skills, such as **brainstorming** and cause-and-effect analysis.

A second cause of delay and failure of SPC programs that are started for purposes of satisfying customer demand is the resistance by the local work force to a tool they perceive, and rightly so, as another management window-dressing exercise. In order for SPC to be fully and successfully implemented it must be part of an overall and formal continuous improvement effort. This continuous improvement effort must reach into every aspect of the company and reflect the culture of the organization. To be more competitive in today's world marketplace, many American companies have to accomplish nothing less than changing their corporate culture.

In our story, Joe was not only implementing SPC; he was also changing the culture of the organization. Operators were being brought into problem-solving exercises, and people were beginning to solve problems by collecting and analyzing data. The introduction of control charts on the shop floor will eventually make any type of random sample inspection plan obsolete, and the control charts will give the responsibility for quality back to the operator, where it belongs. The very fact that Joe was allowed the time and the human resources to properly investigate a problem, analyze the measurement systems, and collect and analyze process data indicates a basic change in the attitude of TW&C management from the more traditional management quick-fix approach to problem solving.

Of course, this is a story, and Joe was able to accomplish so much because it is a story. Joe was able to initiate a basic change to the corporate culture with the tacit approval of the president and his staff. In real life, change such as was brought about by Joe is seldom, if ever, brought about by middle management; sustained change is only brought about when upper management decides it is necessary. And it takes more than tacit approval by upper management to bring about substantial change.

American management has a history of offering tacit approval to programs that are supposed to cure a number of corporate ills. Quality Circles immediately come to mind. The Quality Circle was a concept whereby volunteer groups of workers would band together on their own time in order to identify and solve problems, thereby improving quality and productivity in their areas. Quality Circles were a reasonably short-lived experiment in the United States, primarily because no problem-solving tools or group dynamic techniques were offered to the circles of interested volunteers.

Zero Defects is another management panacea that has come and gone. Under this concept workers were to take special care not to make any defective material. This presupposes the idea that before the implementation of a Zero Defects program, workers were not properly motivated to make products without defects. This basic problem with the Zero Defects philosophy is well-stated by Juran and Gryna in *Quality Planning and Analysis*.[8]

> The fatal assumptions of the ZD movement were (1) that operators are the main source of quality troubles, and (2) that all operator errors can be remedied by proper moti-

vation. These assumptions found favor with some managers but had no basis in fact. What the facts show is that: The bulk of defects (over 80 percent) are management-controllable, not operator-controllable.

At the moment, the new buzzword in and amongst American companies is SPC, and as in the cases of Quality Circles and Zero Defect programs, some American management groups are demanding that their companies have SPC in order to maintain old and capture new business. What these groups of American managers don't realize is that they are asking the impossible. They are asking a very limited number of people within their organization to initiate some very basic changes to the company culture without the complete support of upper management. Worse yet, these basic changes are expected from the entire organization below the corner office, while upper management goes about the usual business of concentrating on short-term gains.

Statistical process control can only be fully implemented if it is recognized by upper management, at the start, as being *a useful tool* with which to understand and reduce variation. Statistical process control must be recognized by upper management as *one of several tools* with which to achieve continuous improvement in every aspect of the business. Management must recognize SPC as being *a common language* with which the local work force can demonstrate to management what they know of the process. Accordingly, management must be willing to provide a sufficient amount of SPC training to the local work force. Management must also create an environment that is conducive to the local work force learning and applying their newfound skills. If operators, supervisors, and lead people view SPC as a club with which management will beat them over the head, they will not be inclined to help fashion the club. Finally, management must offer more than tacit approval to an SPC effort as part of an overall continuous improvement effort.

The best possible way for management to be proactive

in the area of continuous improvement is through the formation of a Continuous Improvement Steering Committee, with the following responsibilities:

Problem Identification—An initial brainstorming session is the best means of quickly and effectively achieving consensus regarding the most important problems facing an organization.

Selection of Project Teams—Dr. Joseph M. Juran defines a project as a problem scheduled for solution. Members of the Steering Committee are best equipped to select knowledgeable members of the organization to work on specific problems. They are also in the best position to moderate the project team approach to problem solving and to avoid the danger of overteaming.

Monitoring Project Team Progress—It is vital to the success of the continuous improvement effort that project teams remain dynamic and not get bogged down. The best way to monitor project team progress is for the Steering Committee to read published minutes of the periodic project team meetings and to ask appropriate questions of the team members.

Implementing Team Recommendations—Only the Steering Committee, which consists of the organization's policy makers, can properly evaluate and reject or accept and implement a project team's recommendation. Written communication from the Steering Committee to the project team regarding the final resolution and implementation of their recommendations is a key ingredient to the continued success of the project team activity.

Holding the Gains: It is very important to make certain that valid, valuable team recommendations, once implemented, do not backslide. The Steering Committee can insure that the gains are held by meeting annually to review final reports and score continued and sustained progress.

Maintaining the Momentum—A successful continuous improvement effort rests on qualified project team facilitators—people who, in addition to their regular duties, understand enough about basic SPC and problem solving to guide project teams through the problem-solving exercise. The Steering Committee can best identify and provide the training for project team facilitators. Momentum can also be maintained through a system of recognition newsletters, bulletin board announcements, and other highly visible management-sponsored activities.

American management must rise to the competitive challenge of the 1990s and fully implement statistical process control as part of an overall continuous improvement effort. The urgency has never been greater.

Dr. Deming, the individual most often recognized as being primarily responsible for the miracle of Japanese productivity and quality, has this to say regarding the implementation of SPC and continuous improvement: "It does not matter when you start just so you begin at once."[9]

Appendix of Formulas and Factors

FORMULAS

Average $(\overline{X})$ = X total (sum of data points in subgroup) divided by number of data points in subgroup

Range (R) = Largest measurement in subgroup minus smallest measurement in subgroup

Grand Average $(\overline{\overline{X}})$ = Summation of averages divided by number of averages

$$\frac{\overline{X}_1 + \overline{X}_2 + \overline{X}_3 + \overline{X}_4 + \ldots + \overline{X}_n}{n}$$

Upper and Lower Control Limits (UCL and LCL) for X Bar and R Chart:

$$\text{UCL for } \overline{X} = \overline{\overline{X}} + (A_2 \times \overline{R})$$
$$\text{LCL for } \overline{X} = \overline{\overline{X}} - (A_2 \times \overline{R})$$

$$\text{UCL for } R = \overline{R} \times D_4$$
$$\text{LCL for } R = \overline{R} \times D_3$$

Upper and Lower Control Limits (UCL and LCL) for Individual Control Chart (note—sample size equals 2):

$$\text{UCL} = \overline{X} + [3 \times (\overline{R} \div d_2)]$$
$$\text{LCL} = \overline{X} - [3 \times (\overline{R} \div d_2)]$$

Percent defective = Number of defects divided by total inspected

CONTROL CHART FACTORS

Number of Samples in Subgroup (n)	A_2	D_3	D_4	d_2
2	1.880	0.0	3.267	1.128
3	1.023	0.0	2.574	1.693
4	.729	0.0	2.282	2.059
5	.577	0.0	2.114	2.326

Notes

1. "Survey of Current Business Conditions in the Wire Industry," *Wire Industry News* 16:12 (July 26, 1989): 4.
2. Greenhouse, Steven, "Revving Up the American Factory," *New York Times* (January 11, 1987).
3. Greenhouse, "Revving Up the American Factory."
4. Miller, William H., "U.S. Manufacturing on Whose Turf?" *Industry Week* (September 5, 1988): 53.
5. Deming, W. Edwards, *Roadmap for Change, The Deming Approach.* Videotape.
6. Juran, Joseph M., *Juran on Quality Improvement Workbook* (Wilton, CT: Juran Enterprises, Inc., 1981), pp. 1–2.
7. Blustone, Mimi, Zachary Schiller, and Otis Port, "Special Report: The Push for Quality," *Business Week* (June 8, 1987): 131.
8. Juran, Joseph M., and Frank M. Gryna, Jr., *Quality Planning and Analysis: From Product Development Through Use* (New York: McGraw-Hill, 1980).
9. Deming, *Roadmap for Change, The Deming Approach.* Videotape.

Glossary

Accuracy. Comparison to a known standard. A measurement device that measures a known standard without appreciable error is said to be *accurate*.

Assignable cause. An unusual occurrence that takes place within a process; unusual in that the variation resulting from an *assignable cause* is outside normal variation. Operators can usually detect and remove *assignable causes*.

Attribute data. Counts or quantitative data that have only two conditions (good or bad, acceptable or rejectable, etc.). *Attribute data* are whole numbers; for example, number of rejects.

Average ($\overline{X}$). The mean value of a subgroup. The *average* is calculated by adding the values within the subgroup and dividing by the number of values.

Brainstorming. An exercise in which a group of individuals express, and record, ideas concerning a specific problem. The essence of such an exercise lies in the openness and creativity of a free and unbiased exchange of ideas.

C chart. A type of attribute chart that helps monitor the number of defects in a standardized unit of inspection. For example, the number of defects in each assembled automobile could be better understood and controlled through the use of a *C chart.*

Capability. The ability of a process to produce all product within customer specification. A process is said to be *capable* when it meets the minimum requirement of having the process average at nominal and taking up no more than 80% of the tolerance.

Cause-and-effect exercise. A problem-solving tool used to identify the root cause of a problem. Fishbone diagrams and brainstorming are very effective means used to identify the root cause.

Continuous improvement. The management philosophy that embraces the concept that processes must be continuously evaluated and improvements made to reduce rework, scrap, and other forms of waste. The goal of *continuous improvement* is to make an organization more competitive.

Control chart. A statistical process control tool that initially helps define the stability and the capability of a process. Production *control charts* are best applied to maintain stability, identify assignable causes, and reduce overall variation.

Control limits. Boundaries on control charts that identify the amount of normal variation we can expect from a process.

Discrimination. The ability of a measurement device to detect sample variation. A commercial twelve-inch ruler would not have enough *discrimination* to detect the variation in length within a box of ballpoint pens.

Grand average ($\overline{\overline{X}}$). The average of all the averages. The *grand average* represents the central tendency of the process.

Histogram. A graphic tool that illustrates, in bar chart form, the distribution of data.

Individual control chart. A form of control chart developed from individual variable points of data rather than averages of subgroups. An *individual control chart* is very useful in continuous processes such as extrusion.

Instability. The state that results in a process when assignable causes are present.

Japanese-style management. The style of management that incorporates the continuous improvement philosophy and employee involvement in the improvement of processes.

Lower control limit. The lower boundary of a control chart; *See also* **control limits.**

Measurement error analysis. The statistical study of a measurement process to determine if its accuracy, precision, and discrimination are adequate.

Natural process limits. The bounds of normal variation inherent within a process. *Natural process limits* are graphically represented by the extremes of the normal curve.

Nominal. The midpoint of a specification. In general, the *nominal* specification is the most desirable to achieve.

Normal curve. Sometimes called the bell curve because of its shape. The *normal curve* is characterized by having one peak and symmetrically trailing off to extreme ends on either side of the middle. A *normal curve* results when variable data is plotted from a process operating without any assignable causes present—only normal variation.

Normal/common variation. Variation due to natural and random occurrences. The differences of height among high-school freshmen is a good example of *normal variation.*

P chart. A type of attribute chart that helps control the percent of defective product yielded by a process.

Pareto chart. A tool that displays in bar chart form a frequency of occurrence within certain categories of interest. Usually the highest category appears in the extreme left and subsequent categories are presented in descending order.

Pareto principle. The principle that in any population which contributes to a common effect, relatively few of the contributors account for the bulk of the effect.

Percent defective. The percentage of a population that does not conform to specification.

Population. A term that refers to the total of everything that is under discussion or study.

Precision. *See* **Repeatability.**

Probability. The likelihood of occurrence of a particular event.

Process. Any combination of people, equipment, raw materials, and methods formed to complete an assigned task of manufacturing or service. Cabling is a *process,* as is answering a switchboard.

Process average. The central tendency of a process. The *process average* is determined by collecting and converting data to a form such as a histogram or control chart.

Process capability study. An analysis of the ability of a process to produce product within specification. In order for a process to be capable it must first be stable.

Project team concept. A project is best defined as a problem scheduled for solution. A *project team* is a group of knowledgeable operators, supervisors, engineers, etc., working together to solve a specific problem. *Project teams* rely upon data collection and analysis, brainstorming, cause-and-effect analysis, etc., to properly address and solve the problem.

Range (R). The arithmetic difference between the largest and the smallest value within a group of variable data.

Repeatability. The ability of a measurement system to record the same value within a series of measurements taken on the same sample, by the same person, using the same instrument. *Repeatability* is also called **precision.**

Sample size (n). Statistical process control is based on the concept that the quality of an entire process can be confidently determined through the statistical evaluation of groups of random samples. The convention for identifying the subgroup *sample size* is through the letter "n." For instance, $n = 5$ indicates that a subgroup size of five is required.

Stability. The ability of a process to produce consistent and predictable output. A *stable* process has only normal/common variation present; no assignable cause variation is present.

Standard deviation. A means of measuring the width of a distribution. A normal curve has three *standard deviations* on either side of the process average. The term sigma is often used interchangeably with *standard deviation.*

Statistical process control. A term used to describe the tools and techniques that quantify and help reduce the variation within processes.

Statistical product control. A term used by the author to describe a misuse of control charts. When statistical process control charts are used to monitor product and make adjustments to compensate for periods of process instability, *statistical product control* is being applied.

Tolerance. The amount of allowable variation about the nominal. *Tolerances* are often imposed without any quantitative knowledge of the capability of the process.

Uncommon variation. Abnormal variation due to some assignable cause.

Upper control limit. The upper boundary of a control chart. *See also* **control limits.**

Variable data. Data that is measurable on a continuous and incremental scale. The length of a rod, the diameter of a wire, and the elongation of a material are all examples of *variable data.*

Variation. Measurable deviation about a known point.

X bar and R charts. Statistical process control charts that help measure and control the amount of variation within a process through the analysis of variation within as well as between subgroups.

Zero defects. The concept that if workers are properly motivated they will make no errors.

Index

Extruder
  microscope, 78, 80, 84-86
  startup scrap study by, 95-98
Extrusion line, Pareto chart
  maintenance on, 112

Facsimile (Fax) equipment,
  10-11
Factors for determining con-
  trol limits, 40
Ford Motor Company, 11, 130
France, as colonial power, 3

GE Consumer Electronics
  Group, 2
General Electric Company, 2
Germany, as colonial power,
  3
Grand average, calculation of,
  39

Harley Davidson, 11
Histograms, 15-19, 74

Identification of problem, 97-
  98, 134
Imports
  quality of Japanese, 8-11
  as threat to U.S. industry,
  1-3
Improvement. *See* Continuous
  improvement; Measur-
  able improvement

Incoming inspection, 53-61
  calculating average range,
  57
  calculating averages, 54-55
  calculating ranges, 56
  collecting samples and
  measurements, 54
Inconsistency. *See* Unstable
  process
Individual control chart, 46-
  51, 53-65
  applications, 49-50, 61-65,
  89
  and decision making, 63
  limits, 50
  measurement error analysis, 74-86
  and problem solving, 62-63
  process capability study,
  106-8
  and quality/productivity
  improvement, 63-64
  range determination be-
  tween individual read-
  ings, 46-47
Inland Steel, 11
Inspection
  for defective products, 7-8,
  116
  individual control chart for
  incoming, 53-61
  100 percent, 121-22
  100 percent versus SPC,
  116
  random sample obsoles-
  cence, 131
Instability
  determination in individual
  control charts, 50-51,
  61

BOOKS AVAILABLE FROM QUALITY RESOURCES

Benchmarking: The Search for Industry Best Practices That Lead to Superior Performance ▪ Robert C. Camp
320 pp., 1989. / ISBN 0-527-91635-8 / Order No. 0187

Company-Wide Total Quality Control ▪ Shigeru Mizuno
313 pp., 1988. / ISBN 92-833-1100-0 / Order No. 0027

Designing for Quality: An Introduction to the Best of Taguchi and Western Methods of Statistical Experimental Design ▪ Robert H. Lochner and Joseph E. Matar
254 pp., 1990. / ISBN 0-527-91633-1 / Order No. 0189

Guide to Quality Control ▪ Dr. Kaoru Ishikawa
225 pp., 1982. / ISBN 92-833-1036-5 / Order No. 0002

Human Relations—The Key to Quality ▪ Horst U. Lammermeyr
296 pp., 1989. / ISBN 0-527-91628-5 / Order No. 0168

Introduction to Quality Engineering: Designing Quality into Products and Processes ▪ Genichi Taguchi
200 pp., 1986. / ISBN 92-833-1084-5 / Order No. 0004

Quality Service Pays: Six Keys to Success! ▪ Henry L. Lefevre
370 pp., 1989. / ISBN 0-527-91629-3 / Order No. 0167

The Quest for Quality in Services ▪ Dr. A. C. Rosander
585 pp., 1989. / ISBN 0-527-91644-7 / Order No. 0188

SPC Simplified: Practical Steps to Quality ▪ Robert T. Amsden, Howard E. Butler, and Davida M. Amsden
300 pp., 1986. / ISBN 0-527-91617-X / Order No. 0146

Statistical Methods for Quality Improvement ▪ Edited by Hitoshi Kume
231 pp., 1987. / ISBN 4-906224-34-2 / Order No. 0147

System of Experimental Design: Engineering Methods to Optimize Quality and Minimize Costs ▪ Genichi Taguchi
2 vols., 1176 pp., 1987. / ISBN 0-527-91621-8 / Order No. 0150

Teamwork: Involving People in Quality and Productivity Improvement ▪ Charles A. Aubrey, II and Patricia K. Felkins
180 pp., 1988. / ISBN 0-527-91626-9 / Order No. 0165

QUALITY RESOURCES
A Division of The Kraus Organization Ltd.
One Water Street, White Plains, NY 10601
800-247-8519 or 914-761-9600